M000317194

Adventures in Urban Bike Farming

Kollibri terre Sonnenblume

Kollibri terre Sonnenblume

Macska Moksha Press
Portland, Oregon, USA

Copyright 2015 by Kollibri terre Sonnenblume

All rights reserved.

Front cover photo copyright 2009 by Calliope Star; cover design by the author.

Published in 2015 by Macska Moksha Press
9035 SE Washington Street
Portland, Oregon 97216
macskamoksha.com

Library of Congress Cataloging-in-Publication Data
Sonnenblume, Kollibri terre, 1969-
Adventures in Urban Bike Farming / Kollibri terre Sonnenblume
ISBN: 978-0-9861881-3-8
1. Agriculture 2. Bicycling 3. Culture and cultural processes I. Sonnen-
blume, Kollibri terre, 1969-. II. Title

For Elaine Close

Table of Contents

Introduction

"When a well-packaged web of lies has been sold gradually to the masses over generations, the truth will seem utterly preposterous and its speaker a raving lunatic." (Dresden James)

This book is about my experiences as bicycle-based urban farmer in Portland, Oregon, USA, from 2004 through 2010 (with a brief reprise in 2013). The arc of the narrative begins with wide-eyed idealism and ends with disillusioned acceptance. If you're looking for a message of "rah rah, look how sustainable we are!" you won't find it here. If, on the other hand, you are genuinely concerned about our collective future and are interested in how urban farming could play a part, there are lessons here to glean.

Mainly, this book tells the story of "Sunroot Gardens," an agricultural experiment that I created and directed from 2007-2010. Sunroot, as it was called for short, took the form of a "CSA," which stands for "Community Supported Agriculture." CSA is a business model in which customers pay a lump sum of money to a farmer at the beginning of the agricultural season in exchange for a share of fresh produce distributed regularly throughout the season. The idea of CSA was invented in Japan and started gaining popularity in the USA in the early 2000's. For the farmer, it advantageous because that a certain amount of monetary income is guaranteed. For the customer, a close and even personal relationship with a farmer can be enjoyed, as well as fresh, local food—*very* local in the case of urban farming.

How is "urban farming" different from "gardening"?

The majority of people who garden enjoy the activity as a hobby but are in no way dependent on it for their diet. The most ambitious ones might cut their summer produce bill significantly and put by an impressive amount of preserves for winter, but they are exceptional, and what they are doing is still not farming. Farmers are trying to provide for themselves by providing for other people, and to succeed, what they provide must be substantial. It is a matter of both scale and seriousness. With Sunroot Gardens, we were earnestly trying to provide for ourselves by providing for other people and at the peak of the operation we had over three acres in cultivation and were providing for over two dozen households. That's why we felt justified in calling ourselves "urban farmers" and in describing our activity as "urban farming."

There were (and are) home gardeners who like to call themselves "urban farmers" or refer to their yards as "mini-farms." While that might be cute, it is trivializes real farmers. Re-localizing our agricultural production system is a real need that will require real effort and real urban farming was what we were seeking to accomplish through Sunroot. For a handful of years we managed to do it, but in the end we found that our efforts could not be maintained, not for logistical or financial reasons, but for social ones.

A particular place in a particular time

Portland, Oregon, between 2004 and 2010 was in transition. Of course, no city is static, and each one is always between phases that are often accurately seen only in retrospect, but the "City of Roses" in that period *was* a particular place in a particular time that made it especially fertile ground (no pun intended) for the urban farming experiment known as Sunroot Gardens.

Passing away was "Little Beirut," a city given that nickname by George H. W. Bush's advisors in the early 1990's because of the energetic protests the President faced there. This Portland was a center of unabashedly leftist politics but was also a homely, low-rent backwater that was perennially overshadowed by its more urbane and glamorous siblings, Seattle to the north, and San Francisco to the south. "Little Beirut" was the city I hoped to find when I moved to Portland in early 2001, hot on the heels of the explosive anti-WTO protests in Seattle in late 1999. I immersed myself in political activism, primarily Indymedia and forest defense. Indymedia was a global network of autonomous alternative media centers based in different cities, including Portland, where the atmosphere led to a special dynamism. During these years, I met and collaborated with hundreds of people, many of whom became supporters of my urban farming in part due to the reputation I had earned as dedicated and hard working.

As is told in the chapter, "The Call of Katrina," I ended up switching my focus from Indymedia to agriculture in 2005. In 2007, the U.S. economy began imploding, though the "Great Recession" wasn't declared until 2008. Over the next couple years, financial institutions failed, gas prices hit an all-time high, and Climate Change crept into the news. Against this backdrop of the system's suddenly apparent fragility—and perhaps even imminent collapse—people in Portland were open to new ideas and big concepts. It was within this milieu, and in directly-stated response to it, that I founded Sunroot Gardens.

As is told in the chapter, "The Media Blitz," local publications from newsprint entertainment weeklies to glossy lifestyle magazines took an interest in Sunroot and other, similar projects, and made a meme of "urban farming." People in Portland who were hungry for creative solutions —especially ones with the stamp of "cool"—proved enthusiastic to lend their help. Resources flowed in, in the form of money, land, labor and more. Thanks to corporate media, I was given what I needed to put my ideas into action. Ironic, to be sure, given that Indymedia existed in opposition to the corporate media, but I accepted it nonetheless.

But by 2010, "urban farming" as a meme had lost its luster and the practice itself had been relegated to the role of one more quirky thing that existed to "keep Portland weird." It was assumed, incorrectly, that urban farming had become an established, successful endeavor and that no more attention needed to be paid to it. This was the final year of Sunroot's operations.

The Portland that emerged next was "Portlandia," a caricature of itself, a destination no longer for scrappy activists—or starving artist, their sometimes partners-in-crime—but for the app-driven digerati, with their oh-so-refined tastes and non-confrontational blue-state politics. Rents skyrocketed, hipsters pushed out hippies, and by 2015, Portland was the most quickly gentrifying city in the USA In short, no longer a hospitable place for unconventional experiments.

In retrospect, it is clear that Sunroot Gardens took full advantage of this particular time and place for as long it lasted, neither arriving too early nor leaving too late (like an engaging work of fiction).

About this book

Stylistically, this book is equal parts historical document, confessional memoir and social critique. Most of the text is quoted directly from the voluminous amount of material I composed at the time, most of it posted to the Sunroot Gardens email list, whose subscribers were CSA share-holders, volunteers, landlenders, and other friends of the farm. I was a very active writer during those years. How active? From the first message, transmitted on April 24, 2007, to the last, on December 10, 2010, the total number of words I sent out totaled just over 100,000. Much of it was routine, relating to CSA pick-up times, work-party announcements, and other logistics, but a significant chunk focused on big picture issues and personal reflections.

I used a very light touch in editing this primary source material. Other than changing names (for reasons noted below), I did nothing except correct spelling and standardize punctuation, and only so the reader wont be distracted by mistakes. Otherwise, every excerpt is intact, as originally composed and sent out into the world, "warts and all." By preserving the text as-is, my intention was to lend a legitimate *vérité* to the story of the evolution of my farming efforts, political awareness and personal growth.

The chapters in the book follow chronological order with the exceptions of the "In Depth" sections, which dig deeper into particular topics —the gardens, the landlenders, and the volunteer scene, for example— and span the entire time-line in their scope.

Apology

Sunroot Gardens was often accused of being a "cat-worshipping cult." We never confirmed nor denied these charges, but it is true that cats played a very special role for us. While biking from one garden to another, if we spotted a cat sitting on a fence post, lying on a sidewalk, or hiding under a car, we would always stop to offer Catnip and see if we could pay tribute with an ear-scratching or a belly-rub. It didn't matter if we were "late" or "in a hurry"; in our opinion, spending time with a cat was the best possible way to spend time, and those moments existed independently of the clock.

Therefore, I issue an apology that this book does not speak more than it does about the many felines who graced our lives during these adventures. Giving them their proper due would necessitate doubling the number of pages in this volume. That this amount of attention, though rightfully deserved, might be tedious to the more boorish readers out there is not the reason for the omission; rather, simple economics at this time preclude me from producing a project of that length. I will take solace in the fact that the cats themselves don't care about whether they are in this book, but I must express my most sincere regret to the humans who are missing out. May Bastet forgive me!

Disclaimer

The events in this book are all "true": that is, they are factual, about real people and real places. Nonetheless, I have changed the names of most people and some places. This is not to avoid being sued for slander; I'm not lying, so there's no libel. Rather, I want the reader to focus on the "big picture" of the social factors that I am illustrating, not on the particular personalities who happened to express them. Readers who were present for any or all of the events therein are invited to enjoy themselves trying to guess who is who. Maybe one of them is you!

Acknowledgements

Thanks to Deva for financial help with publication, to Nikki Hill for personal support while writing, and to Mrs. K for cooking such delicious and healthy meals for me throughout all of it (I couldn't ask for a better wife, dear!). Thanks also to the pre-publication readers who offered valuable feedback: Bryce, James Dunne and especially Laurie Troeger Milliard. Tom Knaust's eagle eye was invaluable in catching spelling and grammar mistakes. Last but not least, I thank the Special Financial Supporters of the First Print-Run: Susan and Ira of Dora's Garden; Tom, Formerly of Mall56; and, doubly, Justin and Jan Stolen.

05.27.2004: "A porch garden in Portland produces delicious bounty"

My first agricultural efforts in the city of Portland, in 2004, were mostly limited to the front porch of the house I shared with Angel, my Indymedia partner and best friend. It was an impressive garden, for what it was. The porch itself was built over the roof of a single car garage so it was bigger than average. Always interested in spreading good ideas and inspiring others to action, I posted an article about the effort to Portland Indymedia website. As was usual for those days, I wrote under a pseudonym.

The joys of urban gardening:
A porch garden in Portland produces delicious bounty
author: Johnny Tomatoseed
27.May.2004 00:23

Raising your own food has many advantages. Rising gas prices are driving up the cost of food in the markets. The insidious spread of genetic engineering is reducing the safety of that food. The industrialization of agriculture funnels increasing amounts of money into corporate hands, rather than to farmers.

These are economic, health-conscious, and political reasons. Less quantifiable but no less real are the emotional and spiritual rewards attained from nurturing seeds or starts into full grown plants that end up on your plate.

Urban gardening presents its own challenges: space, light, uncooperative landlords, etc. I want to share with indymedia readers a few photos and thoughts from my own attempt to grow as much food as possible on a front porch in the middle of the city.

1

These photos are of my porch garden. My landlord allows me only one small raised plot (about 4x8) in the backyard because he wants the rest for his kids and dogs. This is a compromise that i'm willing to make because my landlord actually lives in the same building, so we're sharing this space together. I have no patience, on the other hand, for absentee landlords who won't allow their renters to garden. That's ridiculous. In many other times and societies, it would be criminal to forbid someone to raise their own food. But that's a subject for a different post (or for comments to this one).

So, since that small plot isn't much space, I started a container garden on the front porch to supplement it. Over the course of several months, I collected 5 gallon plastic buckets from a co-op. I drilled holes in the bottom, filled them with soil I bought from local organic farmers (which was much less expensive than buying soil by the bag), and mixed in casings from a worm composting bin we'd been maintaining in our kitchen for the 9 months previous. I tied strings from the bucket handles to the top of the porch roof, and planted starts in them: snap peas, collards, mizuna, spinach, and red dandelion greens. As the peas grew, they climbed the strings. The greens were happy growing lower down in each of the pots. Within 6 weeks I was harvesting greens, and within 8, peas.

The greens have all started to bolt now. (That's when they send up a long tall shoot from the middle and flower and go to seed. They also stop producing many edible leaves at that point, and the flavor of the remaining ones often changes for the worse.) But the peas are in full-on glory. I can go out there and pick a handful every day. I suppose you could make a great stir-fry with peas like these, but I always just eat them right there, sitting on the bench among all the greenery.

The space has become an almost magically pleasant one to hang out in. The dappled sunlight shimmering through the pea vines is quite beautiful, and it is cool on hot days. Bees and other insects buzz around, and there's even a few worms in some of the buckets. When I water the buckets, the rich smell of wet earth hangs in the air with a sultry presence. The space is both productive and alluring; it nurtures and satiates and enlivens. It has become a tiny oasis in a sea of concrete.

This kind of gardening is not difficult. It just requires dedication.

And the tangible and unquantifiable rewards are obvious. My life is definitely better for having this garden in my life.

When the peas are done, cucumbers will take their place, for pickling. i've started those already, from seeds, in another set of buckets, and when they are ready to start climbing, the peas will be about done and the buckets can be swapped out. Here in Cascadia, our long growing season means that you can have two or even three generations of a crop. As food prices rise and global warming makes other parts of the world less habitable, our position here will be a good one. The time is now to start learning self-sufficiency skills like these.

Looking back at this article, I am struck by how well-informed I had already become about the state of agriculture in the world and by my awareness of "the emotional and spiritual rewards" associated with growing food. These dual focuses—the political and the personal (the "spiritual")—would remain co-present and active during my entire urban farming stint, deepening with time.

The Call of Katrina

"The future's here, we're it, we're on our own."
(John Barlow and Robert Weir)

August 2005: In the days leading up to Hurricane Katrina's landfall, widely-publicized forecasts were predicting a direct hit on New Orleans. I assumed this disaster would be the excuse for much flag-waving as a heavy-duty rescue operation took place. After all, the USA's military was increasingly bogged-down overseas, the mainstream economy was performing poorly, and the resident of 1700 Pennsylvania Avenue was losing popularity. Televised heroics could provide a distraction from all of this and more.

A few days after Katrina carved her swath of destruction across the city and region, flags were certainly being waved as FEMA, the Red Cross, and the other big players arrived, but the response was woefully inadequate, even botched. I was puzzled. After all, this wealthy nation had the means for a massive humanitarian undertaking. Why was it not doing it? Then I realized, with shock, that the drama playing out was no accident: an entire city was simply being cut loose. The system was more brutal than I had believed. I had been naïve.

In the weeks that followed, I had friends who traveled to New Orleans to help out. They were people from Portland's radical cliques: tree-sitters, herbalists, Indymedia activists and anarchists, and they immersed themselves in community efforts at the neighborhood-level. Their stories started to filter to Portland back via email and word-of-mouth. The government agencies and non-profits, being authoritarian by nature, could only apply a top-down approach, which inevitably lacked nuance and true responsiveness. In contrast, the grassroots efforts of my friends—focused on clean-up, medical help, and housing—were making a real positive difference, bringing together the forgotten and abused and empowering them to help themselves. I was impressed.

More than just impressed, I was affected. For me, Hurricane Katrina illustrated that we cannot count on big bureaucracies during times of crisis. Furthermore, I believed that "times of crisis" would become "the new normal" during my lifetime and that the longer we delayed collective preparation, the more difficult it would be later, when the shit really hit the fan. Many areas needed to be covered—housing, medicine, water, food, etc.—so I decided I would "volunteer for food" immediately.

I made this decision public by passing out the major portion of my sweet corn harvest at a Portland Indymedia meeting where I announced that I was dropping out to focus on farming. I had been a serious Indymedia activist for five years, contributing the majority of my waking hours for much of that time, so this was a significant shift. On that day, farming officially moved into the center of my life as my activist path. Fortunately, I was not limited to a front porch anymore. That same Spring, a large plot of land had fallen into my lap.

At the time I was a junkie of the local co-op's farmers' market and I never missed a week. I would arrive early to see what was new (and would sell out quickly) and then return at the end to cut deals for unsold produce that the farmers would rather part with at a discount than drag back home. I eagerly explored every stall, asking not just "what is this?" and "how do you cook it?" but "how is it grown?" The co-op's market was then the only year-round farmers' market in Portland, so I also received an education on winter-harvested vegetables, which had previously been unknown to me, being from the Midwest.

Maks and Libby of Wild Things Farm were two farmers I befriended. Maks especially had taken an interest in my gardening when I was limited to the porch garden. One day in the Spring of 2005, he called me over to his booth excitedly, put a piece of paper with a name and phone number on it in my hand, and announced, "Here's some land for you. Give this guy a call." He flashed me his wide smile and patted me on the back warmly.

Minutes later, I was on the phone with the man and got the details. That afternoon I biked up to have a look and was immediately entranced. The site was a big open lot, over a quarter acre in size, with full sun to the south and the west.

The man introduced me to the owner, Susan, who lived on the hill above the plot; her house overlooked it like a sentinel. (Indeed, when I described the plot to people in the years that followed, they would often ask, "The one with that big house looking down on it?") Susan was a middle-aged Italian woman, strong-willed and vivacious. I put on my best Midwestern manners as I articulated my ideas and I ended up impressing her enough that she granted me the plot the very next day. Over the next two seasons, I found her easy to work with. She only had one rule: no marijuana growing. "Not that I have anything against it personally," she hastened to add, in typical Portland fashion.

I named this project, "Lemon Balm Garden," after the *Melissa officinalis* that volunteered there. I had never see the plant before so to me it was exotic and fascinating, though I later found out it was quite common, even considered "invasive" by some. But I always worked around the Lemon Balm plants and with weeding and watering they bushed out and thrived.

Some people claim that magic fairies live in gardens. I don't know if that's a fact, but one night I saw twinkling lights that I could not explain, deep in a patch of Lemon Balm. They didn't disappear as I approached. I felt all the hairs stand up on the back of my neck as I peered into the thick foliage. What was I seeing? When I stood up, I experienced a sudden head rush and almost fainted. Did I have an encounter with the ethereal or was it just low blood pressure? I will never know for sure, but the way I set up the garden, I was certainly inviting magic.

A quarter acre is 110 feet by 110 feet, which would be a double-lot in Portland. The plant-able space at Lemon Balm was both deeper and wider than that, more like 130 x 175: at least triple-lot sized. Quite a step up from the porch garden! I didn't know what to do so I started out with a purely fanciful design. My friend, Bison, drove a large stick into the ground at the exact center of the lot and then I laid out beds around it in concentric circles, and threaded the paths among them. The result was a labyrinth of spirals, switch-backs and dead-ends. It was purposefully difficult to navigate, a fact I compounded by declaring that the plants had "infinite air rights" so it was prohibited to step over the beds. The idea was that by slowing people down and forcing them to go the long way around, habituated patterns of social training could be broken, or at least f'd with. I don't know about anyone else but it had that effect on me.

With this non-conventional layout I was also aiming to utilize "sacred geometry," an esoteric concept of mysterious origin then floating through Portland New Age circles. Supposedly, by planting in beds of certain shapes that are aligned in particular directions, the plants would be healthier and give higher yields. Who knew, but it certainly couldn't hurt, could it? It turns out it could.

As I would learn at Lemon Balm and other gardens over the coming years, designing first for aesthetics does not encourage plant growth or increase production. Starting with the *needs* of the plants is a better approach, but I honestly didn't know much about those yet. I was fully aware, however, that the project was an experiment and that I should keep my expectations low. I planted tons of seeds and starts and then watered, weeded and watched. Gradually, I began learning.

That first season, I prepped and planted less than half of the area and just mowed the rest of the lot. This was enough for Susan, who primarily wanted someone else to take care of upkeep for her, so I wasn't under any pressure other than my own. This was fortunate because I didn't harvest impressive yields of anything. The learning curve was steep. I wasn't disappointed, though. I spent the summer being outside, getting sun, playing in the mud, eating fresh food straight off the plants, and having a wonderful time. It was my first introduction to the farming lifestyle, and I was hooked. Susan invited me to work the plot again the next season.

In 2006, a change in employment pushed me to commit to farming full-time as a livelihood. I had been working for the local food co-op since 2003 to support myself. I enjoyed the job because it was idealistic and centered on natural foods. But over my tenure there, I watched in disdain as its priorities shifted. Having established itself as a little hippie store in the 70's, it was now shaking off the mellow vibe and down-home community radicalism for an uptight bureaucracy and a slick, corporate-flavored conformism. By taking a stance against these things, in little ways and big, I sided with the old-school crowd—both on the staff and among the customers—who were sincerely dedicated but losing their numbers and their voice.

Unbeknownst to me, a small cabal of managers wanted me out. They were watching me like a hawk and building a case against me. The whole convoluted story is not only long but is peripheral to this book, so suffice it to say that eventually charges were trumped up and I was given an ultimatum: quit or be fired. I refused to quit—I wanted them to work for it— and ended up getting a better severance deal as a result; the meeting to hash out my expulsion was attended by the entire staff and some people had enough affection or sympathy for me to soften the blow. But it was all grotesquely political. The sordid nature of the situation was revealed shortly afterwards when my old position was filled by a woman who was having a surreptitious affair with one of the cabal managers. It was like high school, but as I have observed, you shouldn't expect a higher level of maturity from most adults.

As it turns out, food co-ops all across the USA were going through similar changes, shedding their rootsy past in favor of a snooty future. This trend was led in part by national and regional co-op marketing organizations and reflected a general slide towards an increasingly rightward-leaning mainstream that was afflicting everything and everyone fringe in the post-9/11 world.

My own life had been leading me in the opposite direction for several years, away from the center. Being fired from the co-op was like getting kicked off the edge entirely. An astrologer friend pointed out that I happened to be going through my Saturn-squares-Saturn, a major transition period when the old would be ripped away to make room for the new. In reference to the cabal, she said: "They were the agents of your liberation."

However I chose to interpret events, my main source of income was gone and I needed some cash so I could take the 2006 agricultural season seriously. I was in luck: a friend was enthusiastically burning through a trust fund, handing out money left and right, mostly to artists and musicians for all sort of schemes and shenanigans. I sat down with him one day with a bottle of wine, a selection of fancy cheeses, and a detailed spreadsheet. I had made up a budget for seeds, tools, soil amendments, etc., that totaled $5 less than two grand. After my presentation he nodded, smiled, paused for dramatic effect, and then said, "I can take care of *this* part," and pointed at the budget's bottom line. I was quite grateful and ever after he was known as "The King" and received the biggest and the best that I could grow.

With that royal investment, I dove into the 2006 season at Lemon Balm Garden with enthusiasm. I expanded the beds to the edges of the lot. I built trellises for pole beans, set up cages for tomatoes, and double-dug beds for carrots. I seeded lettuce and radishes, kale and turnips, mustard and beets, chicory and parsnips, arugula and sweet potatoes, and much more. A former farmer who lived next door to the garden lent me an irrigation system so I was able to water almost everything with efficient micro-sprinklers that I set up on timers. The place really started to fill out.

By July, I had enough produce to bring some to farmers' markets. I purchased a big aluminum-framed bike cart from a can collector to serve as my farm truck. I was the only person bringing merchandise to the markets by bike, which raised some amused and admiring eyebrows and inspired the word, "cool." For me, doing business by bicycle just happened to be the only way to do it; I didn't have or want a vehicle or driver's license. I wasn't trying to be "cool." But I certainly accepted any sales and all support that came from people who thought I was.

One of the markets I attended weekly was way up in Northeast Portland, started by the neighborhood co-op there. I attended as a favor to the organizer, who was an old friend. The co-op didn't promote it well and it wasn't very lucrative but my ride back down to Southeast took me by The Cash Plant, the party house/recording studio where the King resided with his crazy crew and everyone there loved fresh produce. They bought up all my leftovers at very decent prices and sent me on my way with an equally decent buzz.

My friend Felicia first brought the idea of Community Supported Agriculture (CSA) to my attention, though unwittingly. We knew each other from the Indymedia days and she had been incensed at the way the co-op had treated me (telling a mutual friend that their motive for firing me was that "he wouldn't assimilate") and now she wanted to support my new efforts. "Can you grow me a bunch of greens on a regular basis?" she asked. "I buy them from the store right now but I'd rather give the money to you." I thought that was a great idea and accepted it. I pointed out that this was how CSA worked, but she just shrugged; her offer was about comradeship, not concepts. Another friend who was also upset about the co-op and wanted to stop buying his vegetables there offered me money, too. The two of them became the "charter subscribers" of an embryonic CSA. I tried my best to grow what they wanted and brought them whatever I could every week. These were pleasant social visits that cultivated our bonds. It seemed like a win/win situation to me and I viewed it as practice for starting up a CSA on a larger scale in 2007.

I also sold salad mix to a neighborhood restaurant and had one-off customers here and there, like to folks going to Burning Man. These varied sources of income brought in just enough money for me to get by, but only because my rent was very low. At that time I lived with my friend, Homitsu, who owned the house. When I got fired from the co-op she lowered my rent to $50/month so I could try to make a living by farming.

She, too, was not happy about the co-op. My astrologer friend's interpretation certainly seemed to be ringing true: events and resources were lining up amazingly well for my new line of work, which was also my new lifestyle. Self-employment and farming were providing two more steps outside of the mainstream. and the further out I went, the freer I felt.

On August 5, 2006, I staged a public prank at Lemon Balm Garden that was intended to push me out further still, but in a more personal way. That was the day I got married. I had made preparations for months beforehand, sending out invitations, registering for gifts, lining up a reception hall and band, picking the principals for the wedding party, lining up the services of an ordained minister, etc.

To say that marriage is a major component of U.S. culture is like saying that bedrock is a major component of the earth's crust: Neither the nation nor the planet could exist in their current forms without these underlying masses. I had been raised by church-goers and attended sixteen years of private religious schools, so I felt that I needed to address the issue of marriage for myself somehow. The answer—which was suggested by, of all people, my girlfriend at the time—was to marry myself.

The prank proved popular. A crowd of people showed up and together we all "played wedding." The occasion was half-costume party, half-theater improv. People took on stereotypical roles and made jokes of them that they played to the hilt: the annoying uncle telling racist jokes; the pregnant bridesmaid drinking like a fish; the jealous father-of-the-groom lustily claiming, "I saw her first!" There was even a live rendition of a Carpenter's song ("On Top of the World").

Strange as it might sound, the event had real reverberations for me for years afterwards, even up to the present day. I actually *did* feel like I had dispensed with the institution of marriage for myself and that felt liberating. Publicly, I had made a declaration my position "outside the box." Tellingly, most women were not interested in me after that, which helped keep life free of some particular flavors of heartbreak. And finally, since I was no longer looking for "that special someone," I could focus more intently on my own path.

Gardening at Lemon Balm had been a deeply enjoyable experience but I believed that the next logical step in my pursuits, both external and internal, was to move to the country. There, I could "farm for real" (I thought) and have some peace and quiet (I hoped). I gave notice to Susan and transplanted my perennial herbs (including most of the Lemon Balm) into a small garden near Homitsu's house, which was intended as a temporary nursery until they were moved out of the city. The season closed with feelings of satisfaction and anticipation.

2007 Season: Forced Landing

"Life is what happens to you when you're busy making other plans." (John Lennon)

As the 2007 farm year opened, I had a clear plan.

I had struck a deal with another farmer, Lancia, whose mother owned land in the country near the town of Westfields, outside Portland. Lancia and I had met at the co-op's farmers' market in Northeast where she was also peddling produce. During slow spells (which was most of the time), the two of us talked about farming, where we wanted to go with it, and what inspired us. Lancia was around my age but this was her first season farming. Though I had only started in earnest in 2005 myself, I was a life-long "plant person" who had been raised by gardeners so she saw that she could learn a lot from me. For my part, I found her pleasant and interesting to hang out with; she was a lesbian and a breast-cancer survivor with a double-mastectomy, so she was not "normal"; her brush with death had given her a wider perspective.

We visited her mother's land together in 2006 to pick blackberries for market so I got to see her set-up. She had three acres altogether in some bottom-land next to a wooded watercourse. She didn't need all the space to herself, so together we came up with a plan to split the area between our operations, but to collaborate on planting, cultivating and harvesting, and to share costs like fuel and bulk amendments. We would sell our produce altogether in two markets every week, but alternate who was taking everything in. That way, each of us would make money from two markets but only have to attend one. Last but not least, I could park a trailer out on the land so I had a place to live.

We talked all this out with her mother, who, despite being rather imperious, issued her stamp of approval. She was proud to see her daughter investing herself in the land, and understood that I could be a real asset. The only tense moment came when Lancia's father, who, like me, was originally from Nebraska, asked if I was a Cornhusker football fan. You would have thought I was confessing to pedophilia from the look on his face when I said, "no." I was reminded of why I had left the Midwest. Fortunately he was not the decision-maker so that wasn't a deal-breaker.

By January, the three of us had hammered out the details and written them up so I put out a call offering CSA shares to my activist friends and to Permaculture circles. I sold a ten CSA shares total, at $300 for a half share and $500 for a full share, which were medium-to-low price at the time. I collected most of this up front and bought seeds, tools, soil amendments, books and market supplies.

"Sunroot Gardens" was the name I gave the CSA. The Sunroot (*Helianthus tuberosus*) is a knobby root vegetable better known as the "Jerusalem Artichoke" and is the perennial cousin of the beloved Sunflower (*Helianthus annuus*). It has similar flowers, though they are much smaller. Sunroots are wicked hardy and very tenacious: they are totally frost-tolerant, thrive in heavy clay soils, spread like weeds, and are nearly impossible to remove from a spot once they establish themselves (which only takes one season). They are the only cultivated vegetable that is native to North America. I found all these characteristics inspiring and also just liked the word, "Sunroot," as it brought the concepts of warmth and light with soil: grounded but reaching heavenwards.

In early February, Felicia helped me move a 14-foot residential trailer out to the land, which is to say she drove the truck that hauled it. It was a challenging task and there was some gritting of teeth and a few curses from her lips. (I returned the favor a few years later by driving a moving truck from Portland to Seattle for her, with some curses of my own on that city's narrow, congested streets.) The trailer was a gift from Harry McCormick of Sunbow Farms, in Corvallis, another farmer friend from the co-op market. Harry's career as a farmer dated to the back-to-the-land movement of the 70's. He was one of the pioneers of the organic movement and a founding member of Oregon Tilth. I wanted to absorb some of the mojo of that crowd and considered the trailer to be a talisman of sorts.

At the end of February I started planting. During this period, I spent half of my time in the city and the other half in the country. To get to the farm from the city, I would ride my bike to a Max train stop in downtown

Portland, take it to the end of the line in Hillsboro and then bike another nine miles on country roads. Those were fun jaunts, and the challenge of the journey added to the sweetness of my time out there. I would work until sunset and then sit back with my feet up, pour a glass of home-vinted Dandelion wine and smoke a spliff as I gazed out over the fields until the stars came out. It was quiet by day and dark at night, in marked contrast to the city. Even though my time was full of labor, I felt myself starting to unwind out there in some deeper ways.

Whenever I arrived back in Portland, my first stop would be the old Red and Black Cafe on Division Street where I would order a "Multi-tasker," which is a pint of stout with a shot of espresso in it. The espresso was for bringing me back up to speed with the city, and the stout was to cushion the blow. While in town I would check in with friends and investors, pick up supplies, and plot out the next steps of the project. Everything was humming along well, all "according to plan."

I lived this lifestyle for just six weeks before it came to an abrupt end. Here is how I told the story to my CSA investors:

Subject: CSA update from Kollibri!
04/24/2007 01:51:48 PM PDT

Hello CSA subscribers!

Here's an email update for y'all, about what's going on with this year's produce. I wasn't expecting to send out anything this season with such dramatic news. This mail might read a little like a novel, which is good because it's a little long perhaps, so it better be entertaining, right? Future correspondence will be shorter and less dramatic (I sure hope).

First, the basics. Here's what's been planted so far this year. You, the CSA subscribers, are first in line for all of this. What's left-over will be sold/bartered at farmers' markets and distributed to helpers, friends, etc. Amounts are measured in size of patch (i.e., 10 x 10), or by row-feet (10').

Peas: sweet crunchy snap peas, 3 colors of snow peas - 60 x 60
Roots: 7 kinds of beets, 9 kinds of radishes, 4 kinds of turnips - 75 x 60
Nantes carrots: 150'
Kohlrabi: 2-3 kinds - 90'

Chard, Bright Lights: 69'

Arugula: 63'

Chicory: 57'

Lacinato Kale: 54'

More arugula (diff. seed): 54'

Spinach: 48'

Ruby Orach: 44'

Chickweed: 42'

Edible chrysanthemum: 30'

Omega flax: 5 x 5

Dandelion: 100'

Mixed lettuce: 40'

Spring salad mix: 50'

Mustard greens: 80'

Fava beans: 350'

Mustard greens (diff. seeds): 6 x 10

Poppies, 'breadseed': 30'

Shallots: 45'

Garlic: 75'

Oats: 70 x 20

Brown flax: 35 x 40

Lentils: 240'

Garbanzos: 60'

I was excited to get so much in the ground. And it's all coming up great so far 'cept for the carrots; they might need a little more warmth first. I put them out darn early, honestly. These veggies cover an area about a quarter acre in size, maybe a little bigger. It is the largest single spring planting I have ever done, by far. Yippee!!

DA DA DUM... The Dramatic Part:

That's everything that got planted out on the acreage in West-fields. As many of you will recall, I have been talking about this land, and working with the farmer there, since late last summer. By winter we had struck our deal for the season, which was this: a) $75 lease for the land, paid to the owner ($50/acre, what she charges everyone else), and, b) a deal with the other farmer that we would do two farmers markets together, with each of us just doing one and bringing the others' produce along. A great deal—you get to offer your produce at two places, but only have to go to one of them yourself.

Well, the deal didn't last. After arriving at the land, the owner added a condition: Pay utilities. Okay. That added $350 for the season to the expense side of the budget. I had to rework the numbers in a serious way by cutting back in other areas. Then, after being there for six weeks and having planted all of the above (plus a bunch of medicinal herbs), the owner demands that I purchase a $1,000,000 liability insurance policy. The cost of that (according to a quote from Country Insurance, which the other farmers around here use) is $625-$700.

So, suddenly, the cost of being there had risen from $75 to as much as $1125, with a big chunk of that going to the freaking Insurance Industry, which is just a form of Legalized Extortion, in my opinion, and exactly the sort of B.S. we're trying to get away from by organizing new distribution systems like CSAs in the first place!

This wasn't going to do, obviously, so a new deal was struck. I will not be purchasing liability insurance. I may still harvest what I've planted so far. However, I may not plant anything else, and I can only go to the farm by myself. And everything's gotta be outta there by Aug. 1. I also reduced the utility charge, and will be paying only $100 total.

So, what this means is that alternate land will be needed to plant everything else that's been planned for the year. I am looking in the City only, for ease of transport. Fortunately, three offers of small parcels have already come in, and today I landed a larger parcel (100 x 30). All are within 12 blocks of my house. I still need to come up with a bit more—another 3,000-5,000 square feet or so to get everything in the ground, but I am confident that this will appear. And best of all, I have not fallen behind my planting schedule! Nothing so far is being planted 'late'; it's just going into places that I hadn't been planning to plant it.

But, like, "No Worries, dude":

Ultimately, you as CSA subscribers should not suffer any interruptions-in-service from these changes in where I'm growing food. In fact, some produce will now be fresher as it will be traveling a shorter distance. However, this distributed model could produce some as yet unforeseen logistical issues that might come to your attention, though I will certainly be working to avoid such scenarios.

Also, if I need help with something, y'all will be some of the folks I will be alerting. Among other things, I might need some assistance with transportation from the country to the city from time to time, in the form of borrowing a vehicle, depending on how things go down with Lancia. In our original deal Lancia was going to be providing a truck, but that's more of a question now.

OK, ALREADY, WHEN DO I START GETTING FOOD?!

Originally, I told people "the end of April," which is now upon us. I am now saying "sometime in May." I have never planted a lot of these crops so early, and I didn't take into account that they grow so much more slowly in the early spring than in the late spring. So, it's all taking longer than I thought it would. But it's all definitely coming along, and at some point will become a Flood. AND, no matter when the season starts, it will last at least 27 weeks total, as promised. We'll go through the end of November if we need to.

SO... I will send out another email in the next couple weeks as I get a better idea about when the exact start date will be.

Please call or email with any questions, ideas, leads-on-land, wishes-of-congratulations, or otherwise. I now have a cell phone number: 503.xxx.xxxx. Call anytime.

xoxoxoxo,

Your CSA Farmer,

Kollibri terre Sonnenblume

"This distributed model could produce some as yet unforeseen logistical issues"—ha! That turned out to be quite an understatement. "Unforeseen" would be the watchword over the next four years and the "logistical issues" that emerged ultimately made the enterprise unworkable. At that moment, though, I wasn't thinking big picture, or any further ahead than the next planting date. My planned flight had become a forced landing. It was all I could do to simply not panic.

The scramble for real estate was my top concern. The first person I contacted was Susan to see if she had found someone else to tend the old Lemon Balm Garden plot. She had, but upon hearing of my plight, wanted to return it to me. The new caretakers, however, would hear nothing about it. They were a non-profit organization and they considered the deal done. Apparently they even made some kind of legal threat. Susan apologized to me and wished me well.

Fortunately I had an expansive network of friends and acquaintances in town so when I put out the word for garden spaces, I got some. By the end of that same month, I had already collected several.

Subject: This is your first week of produce!
04/30/2007 08:09:55 PM PDT

...In other news I will be working with Terwilliger Community Farm in SW Portland to plant much of the summer/fall crops. Altogether, I will be tending to SEVEN different plots in the city to grow everything I had planned. With the addition of the quarter acre I get to use at Terwilliger, I am now set for the season and no longer need to find more land. So, the drama is over regarding the loss of the land in Westfields, and everything's all set to grow wonderful food for all you wonderful people this year.

I am super excited to be offering y'all the first food of the season, and suspect that this is going to be the funnest summer of my life!

That optimism wouldn't last, but it helped me through some challenges in the meantime. The "drama" over the loss of the land might have been over, but the effects would be felt for the rest of the season. April is a late date to start over the farming season in the Pacific Northwest.

At the time, seven gardens seemed like a huge number to tend. Little did I know that a year later the total would exceed thirty. There wasn't much I pegged correctly during this transition, but I was right about one thing: it *did* turn out to be one of the funnest summers of my life.

06.19.2007: First Broken Bike Cart

This email shows how rapidly things were moving and changing during my unexpected return to the city. Six weeks previously, I had considered myself all set with seven plots, but here I have already increased the number to eleven.

Date: 06/19/2007 08:01:39 AM PDT
Subject: This week's CSA share

Hello all,

CSA Share #5 will be delivered tomorrow, Wednesday. After this many weeks, it looks like I'm getting the kinks worked out of the logistics of that process. I thank everyone for the patience and understanding they've shown while I've been figuring it all out. I really do enjoy bike delivery and am glad it's been working for the most part.

This week's share will feature peas again, which people really enjoyed last week, plus the last of the turnips, and a new batch of salad greens, these grown in the city. And we'll see what else is ready!

NEWS:

This was an exciting week. The bike cart busted, and busted bad, but has been fixed by a welder friend who works at the PCC Rock Creek campus. This guy is a friend of my roommate and had seen the cart a year or so ago. He's into creative bicycle transportation and took photos of it. He also looked at the hitch and was like, "That's where it's gonna break." And that's where it did, so he knew what to do. Just another example of how this project is "Community Supported"!

It was Squash Planting week. I had gotten starts going in the greenhouse of one of the CSA subscribers and it was time for them to be moved from their little pots. With help from friends, 60 pumpkins were planted at Terwilliger Farm. Two subscribers were out that day to see it, and a third was helping. I'll be growing those out there with as little irrigation as possible, perhaps nearly none. We'll have an event out there when they're ready, perhaps combined with the Farm's annual pumpkin-carving day, for people to pick them up.

Starts of four kinds of melons, three kinds of winter squash, and white pumpkins were planted at the new gardening space at SE 44th near Harrison. This garden is called "The Firepit" because of the large impressive example of such in the middle of the yard. This might be the location of the August CSA party, more details to follow.

Also in the ground this week: Corn, Black Aztec Sweet. So dark purple it's almost black, and a great sweet corn flavor. Planted from seeds I saved myself.

Sprouting: Beans! 11 varieties of snap, in different colors and shapes. Always one of my favorite things to grow, harvest and eat. I am looking forward to sharing these all with you. They are being grown on a plot that has no irrigation available so this will be a dry-gardened crop. I grew some beans like this last year, so know it can be done.

Thriving: Peppers and eggplants, in the greenhouse. They all doubled in size in the first 10 days and their growth continues to be explosive. We have flowers and fruit already. And some aphids, but we'll get them to move along soon here.

All in all, things are going quite well. Eleven plots is a lot to take care of, but it's working out. I am still feeling the effects of the Westfields acreage unexpectedly not working out for the season—there is certainly less of some of what I had planned to plant/harvest—but the recovery from that event has been impressive, in my mind.

Anyway, I'm off to put some zukes in the ground. See y'all soon!

xoxoxo, Your Farmer,

Kollibri

08.20.2007: "Where are the tomatoes already?!"

Tomatoes are, of course, the emblematic vegetable of summertime, grown by over 90% of home-gardeners in the USA Starting on the first hot day of June, customers of CSAs and farmers markets begin asking where the tomatoes are. But in the mild Northwest, the only way to produce tomatoes before August is with the use of artificial lighting, heating pads under the start trays and hoop-houses. All this special equipment is expensive and Sunroot was in no position to make such an investment. Even with all those bells and whistles, mid-August is still early for tomatoes. Regardless, a rising chorus was taking up the chant, loud enough that I felt the need to respond.

A key to re-localizing diets is for consumers to adjust their expectations. As with many emails to follow, here I tried to educate my CSA customers about the challenges of growing hot weather vegetables in Oregon's maritime climate.

> **Date: 08/20/2007 07:45:05 AM PST**
> **Subject: CSA News: "Where are the tomatoes already?!"**
>
> Dear subscribers:
>
> PICK-UP REMINDER:
>
> Until the end of the CSA season (sometime in November), unless announced otherwise, veggie share pick-up will be Saturdays, 2-7, at my house: [address deleted].
>
> FARM NEWS:
>
> This is some cool cloudy weather we've been having! I've been watching it closely, with an eye towards how the veggies like it /

don't like it. The greens love it. (Not just salad greens but also kale, collards, mustard.) Roots like it fine. (Carrots, beets, parsnips.) The beans seem fine. (Harvested ~75 lb. last week!) The zukes are bursting, if a *little* late. The winter squash and pumpkins seem to be moving along on time. Even the corn and melons (both a little tricky this far north) don't seem noticeably behind.

The tomatoes, though, are definitely slow. Slow enough that folks are starting to notice, including Greg & Claire, the best tomato farmers I know, who sell at People's Market, and who've had very little to bring this year. The only farmers at market with tomatoes have been Herman & Lydia Obrist who are growing theirs in a greenhouse. The cherry tomatoes that have ripened so far have been a bit flat, flavor-wise, and not all that juicy. So there's been nothing to bring y'all yet!

The peppers and eggplants also seem behind to me, but I have very little experience growing either, so I don't really know. I've got 'em in a greenhouse, and the plants are HUGE, but producing very few flowers and fruit. Could be the soil I gave them was *too* rich, could be the weather, could be they're not getting pollinated. I pulled a screen off one of the windows the other day in case we need more bees in there. We'll see.

The taters at Terwilliger Farm also seem behind to me. The plants look a little small and many of them are getting eaten by something. Grasshoppers? Black spider mites? It's unclear. Pulling up a few to check showed tubers, but still small. Perhaps they're just slow. I hope to give them a foliar feeding of some compost tea soon in case what they want is more nutrients. "Foliar feeding" is when you spray the leaves directly. It's an Organic Farming technique for pest control and/or fertilizing. A backpack sprayer has been ordered to help perform this task.

We're only just now starting to catch back up from the set-back dealt to the season by The Westfields Fiasco. Essentially, my farming started over in April, with some plots not even located until late June. That's a crazy time to get a CSA going. That's also just how farming goes: it's unpredictable. I'm enjoying myself and I hope everyone else is, too.

Thanks!

In-Depth — The Firepit Garden

Of all the gardens we tended over the years, the Firepit Garden was the most impressive. When I referred to it as "the crown jewel," I was only half-joking. It was a large plot, nearly a quarter acre, centrally located in the Hawthorne District, a neighborhood known for its hippie vibe, but also the turf of old school working class folks and an influx of snooty yuppies. Starting in the Spring of 2008, the Firepit became the Sunroot Gardens HQ and my home. I never met the legal owner of the property; all my dealings were with Lady Quince, the renter-in-charge.

During its peak years, the Firepit Garden was a vibrant marvel, bursting with botanical life, buzzing with insects, and bustling with human activity. A maze of wood-chipped paths led visitors among the wide beds that radiated out around a large, stone-ringed firepit. No lawn remained. Cats came and went freely, partaking from the bountiful Catnip patches.

Over the years, hundreds of different plants grew at the Firepit Garden. Taking a tour from the ground up: below knee-level were root crops and annual greens—Carrots, Radishes, Turnips, Spinach, Mustard—and sprawling Winter Squash vines. Up to waist-high were the Lemon Balms, Bush Beans, Catnips, Tomatoes, Mountain Mints, trellised Cucumbers, Feverfews, Basils, Daturas and Poppies. At shoulder level were the Pole Beans on trellises, Motherworts, Mugworts, Klip Daggas, Scarlet Mallows and bolting greens. Towering over everyone and everything else were the Sunflowers, Mulleins, flowering Parsnips, Tobacco, and (of course) Sunroots.

Children loved the garden, which was my intention. There were all sorts of nooks for hiding, where you could see other people and they couldn't see you. Paths made of stepping stones invited—and received—playful skipping or careful hopping. Berry bushes dangled their delicious treats at the perfect picking height. Parents would warn the children not to hurt anything, but were invariably the clumsy culprits themselves. The youngsters intuitively perceived that the place was for them and did no harm as they explored its verdant treasures. Novelist, Tom Robbins, once wrote, "It's never too late to have a happy childhood," and perhaps that's what I was trying to do for myself there.

When I first started work at the Firepit in the summer of 2007, large sections of the yard were choked under Blackberry brambles and Bindweed. Blackberries are thorny plants with long roots and a fighting spirit. They're hard work, but that's all. Bindweed is a hardy perennial in the morning glory family, known for its runaway growth, strangling vines and seemingly supernatural tenacity. Its white, trumpet-shaped flowers are misleadingly charming; Bindweed will take over a garden space, choking out every other plant up to the size of a tree sapling. Its deep, extensive root system is notoriously challenging to remove. A piece of root as small as a 1/2 inch long can re-sprout and start a whole new plant if it remains in the soil.

I heard a story about a man who applied Round-Up herbicide to the Bindweed on his property. Not only did the plants in his yard yellow-up and die, but so did the ones across the street. Impressive: the colony had grown under two lanes of pavement and come up on the other side.

I didn't want anything to do with Round-Up or any other poisons, so I dealt with the Bindweed by digging deeply and painstakingly extricating the root system bit by bit. That first summer, I dug up several root balls, each the size of a grapefruit. I never managed to banish the Bindweed entirely, but it was less pervasive after that.

The Firepit Garden provided many lessons, most of them unexpected. In 2007, for example, we rototilled a wide swath across a large portion of the property, unaware that we were not only spreading a bunch of seeds around but also covering them up nicely; planting them, in other words. In 2008, dozens of species volunteered everywhere, but especially one particular plant in the mint family that I did not recognize. I suspected that it was Catnip so I sought a positive ID from an expert: the fe-line next door. His enthusiastic reaction was clear confirmation. I nurtured this accidental sowing that season and in all the following seasons and brought in huge harvests. Many, many cats all over town enjoyed this pungent treat.

Volunteer plants were what made the Firepit. Though every year we would clear and work beds for particular annual vegetable crops, the self-sown population grew in variety and density with each season, until it became difficult to plant anything new there. That was, however, the official policy, inspired not only by the accidental Catnip but by a Japanese farmer friend, "Chabo," who I had met years before at the co-op farmers' market.

In 2005, I visited Chabo's farm and was quite impressed by his practice of "building the seed-bank." Like any serious small-scale farmer—not that there were (or are) many of those—he saved as much of his own seed as possible. But plants that a farmer lets go to seed also drop their seed to the ground and even the best efforts at collection will not prevent escapees; "volunteers" are the inevitable result, which Chabo appreciated. At Chabo's farm, after successive seasons of crop rotation and seed-saving, lush beds of mixed vegetables were coming up on their own every year, leaving him only to thin them out to good spacings.

At the Firepit I took it a step further; when I pruned back bolted tops of seed in the autumn, I would shake them around, whip them on the ground, and strips the pods from the branches as I walked them back to the compost pile. I wanted to see what would do well where—under the apple trees, by the back fence with afternoon shade, on the south side with full sun—so I varied my route regularly to distribute everything evenly around the entire garden.

In the Spring, volunteers popped up thickly everywhere, and so did the learning opportunities. Indeed, certain plants did do better in certain spots, and I saw how sunlight, soil type, and other factors affected them. I also took note of the timing of their germination since that revealed good planting times for sowing them purposefully elsewhere. Occasionally, the germination of a particular plant, acting on its own, was at variance with the local conventional wisdom about it. Parsnips, for example, volunteered in February or March, although most farmers in the area waited until June. I watched as these earliest germinators got the largest with the least care—including less irrigation—because they established themselves thoroughly during the cool, moist Spring and could then fend for themselves when the hot, dry Summer arrived. I discovered that I didn't need to do anything to grow big (and delicious!) Parsnips except let them go to seed and come up on their own and then thin them well.

I haven't heard of anyone else engaging in this method, of "building the seed bank," except for Chabo. I am not surprised at its rarity since it is antithetical to the ordered, planted-in-rows discipline of European agriculture. The garden ended up planting itself, and I would have to move paths every Spring when the volunteers started showing up. Why "weed out" perfectly good plants that are thriving on their own? Because they're in the wrong place? Well then make it the right place! This was the flavor of logic that evolved at the Firepit, and then spread not only through our network of gardens but also through our own evolving decision-making processes.

"Seed-bombs" are a planting technique sometimes used in so-called "guerrilla gardening," in which seeds are combined with mud and compost into little balls and then dried. The ideas is that you can toss them over fences into empty lots and when it rains, the "bombs" will re-moisten and the seeds will germinate. I never tried it—I wanted to personally nurture my plants more than that—but I had a vision of the Firepit Garden as a giant seed-bomb that would eventually spread its botanical treasures throughout the neighborhood in some future time (maybe bereft of humans) when the manicured lawns would give way to a wilder life. That would be something to be proud of!

There was an old single-car garage on the Firepit lot, tucked back from the street on one side, which had seen better days. The roof leaked in multiple locations and you could see through the walls in a few places. Previous renters had attempted creative renovations but these efforts had not stood the test of time. Despite its dilapidated state, the place looked like a greenhouse-waiting-to-happen to me. Lady Quince said the owner had talked about "knocking the thing down," so she doubted if he would mind if we did something with it.

I asked my friend Victory, who was a contractor (among other things), if she thought we could transform it into a greenhouse. She was already familiar with the structure, having known Lady Quince for years (in fact, she was the one who introduced me to Lady Quince in the first place), and in her opinion the project was do-able. She offered her help but said the first step was to acquire some windows.

The next day, I was biking through Inner Southeast in the neighborhood along the railroad tracks south of Powell when, lo and behold, I spotted a stack of windows in the parking strip in front of a warehouse. They were of good quality, in wooden frames, many of them double-paned, and all had the same dimensions, which would make them easy to work with. They had been removed that day to make way for new ones.

The crew was still on site so I asked if the windows were available. They were, I was told, but only until 3:30 when they would be loaded up and hauled to the dump. I checked the time. It was a little after 2:00. I thanked him and hopped on my bike, wondering how I could find quick transportation for my new-found treasure.

Three blocks away, I ran into my friend Tarra with her Volkswagen Van. I informed her about the situation and asked for her help. She was happy to give it—for a price: she wanted two or three windows for herself to make cold-frames at her own house. That was no problem. There would still be two dozen left.

We loaded up the booty right away and brought it up to the Firepit. As soon as Tarra drove off, I was on the phone to Victory. "I've got windows," I told her. "You want to make a greenhouse?" She was astounded to get the word so soon but was happy to honor her offer.

We put together a work crew that included Angel's brother, David G., a talented carpenter and fiddle player. The initial inspection of the garage revealed that most of the wear and tear was superficial; nearly all the structural framing was still in good condition. One exception was the southwest corner: the support beam had rotted away where it met the ground. A closer look seemed to reveal an impossible situation in which the only thing holding up that corner was the roof, but the only thing holding up the roof was that beam. How it was still standing was a mystery. Even though it had made it this far, we propped up a new 4x4 in the spot to make ourselves feel better.

With few carpentry skills myself, I left the details of design and implementation to the crew, and served as assistant, gopher and banker. After a tense meeting in which they reconciled their professional opinions, they got to work.

In just a few weeks, the half-wreck had been transformed into an attractive and well-functioning building that was part-greenhouse, part-barn, and part-residence, with an exterior painted sky blue and Calendula yellow. Most (but not all) of the leaks had been plugged, and I moved into it that Summer (2008). At night, I slept on a comfy pile of burlap coffee bags that I had to roll up and stash away during the day so there would be room to walk around. I ran an extension cord out from the house, plugged in a hot plate and a stereo and strung Christmas lights as the primary source of illumination. A heavy round wooden table served for dining, meetings and projects. I scored metal shelves for storage and put up hooks and brackets for tools. I set up the greenhouse section for propagating plant starts and brought in hoses for irrigation. With its concrete slab floor, exposed beams, and lack of insulation, the place was definitely rustic but totally suited my needs. I was in love with it. I dubbed it "The Fort" (sticking with the children's garden theme) and in the following seasons it played host to innumerable meals, meetings and mischief-making as well as pumping out thousands of plant starts.

Some of the fondest memories of my life are from the Firepit Garden. I was free there, free to experiment and free to grow and I took full advantage of it. My arrangement with Lady Quince was the key factor in this freedom; she gave me complete control over the outside property, setting only two rules: don't build anything "janky" and don't turn on the water when someone's in the shower (because the water pressure was so low that it would reduce the flow in the bathroom to a trickle). She herself held domain over the house, which I rarely entered. Sometimes we would hang out on one of the porches, or make a fire in the firepit, and drink a beer or a cider and shoot the shit.

Over time, the Lady and I formed an enjoyable bond that entered what some people would call the "psychic" realm. Out in the garden, puttering around, a thought would enter my head about something she wanted or needed. The next time I saw her, sure enough, she would bring it up. Or, I would raise the subject first, leading her to say, "I was just going to ask you about that!" I have experienced connections like this with other people over the course of my life, starting with my biological mother, and have always appreciated them, though their mechanism remains a mystery to me.

Without the Firepit Garden, which was both Sunroot HQ and my home, my urban farming would have been far more difficult, logistically speaking. And as a classroom and playground, its value can hardly be overstated; I learned so much there about plants, the seasons and myself. I transformed the space but it equally transformed me.

08.31.2007: "The gophers got the taters"

Date: 08/31/2007 08:47:28 AM PST
Subject: the gophers got the taters

Went out to Terwilliger Community Farm last Sunday and checked on the potatoes. About two months ago their foliage had stopped growing. I wasn't super worried about that since weather- (or what-ever-) related pauses in vegetable growth happen sometimes, but I became concerned when—a couple-three weeks later—the leaves began showing holes (from black spider mites? grasshop-pers?) and then turn yellow-ish. Tater foliage turns yellowish when the roots are ready, but this seemed premature. Every week I checked them. Then they went from stalled-out to declining in the 10 days previous to last Sunday. So we dug 6 of the 10 rows up, which left 3. One row had already been unearthed for the CSA Dinner Party and was definitely a lower-than-expected harvest, but I didn't know whether that was just that particular variety. (I planted seven different varieties altogether.)

The rows were 45-50 feet long each. One could reasonably expect 50-75 pounds of potatoes from each row, even with little-to-no arti-ficial irrigation. (Taters have been dry-farmed since time-out-of-mind in low-rainfall areas in 'the U.S.', to say nothing of their re-gion of origin in the arid Andes Mountains.) However, what we got was 2-5 pounds per row. Yes, you read that right: 2-5 pounds per row. That's what I would term a "spectacular crop failure" (as in, it's a spectacle what a failure it is).

So what happened?

When we dug up the rows, I got down on my hands and knees to check it out thoroughly. Everywhere I found holes and tunnels and little piles of dug-earth. What's that mean?

GOPHERS!

Yup, the gophers got our potatoes. Just about all of them. The remaining three rows don't seem AS affected, so I'll dig those up later and we'll see. But I'm not expecting much.

So, what have I learned from this?

1) Don't plant taters—or any other root crop—at Terwilliger. (The house's vegetable garden there, which is mostly non-root veggies, seems untouched by the gophers.)

2) Now that I recognize the signs of gopher invasion, consider possible courses of action. These include, but are not limited to: traps, planting Euphorbia lathrys (which they don't like), sulfur ignition in the holes (a method that is acceptable under USDA organic standards), training small dogs to get them out, or finding a friend with a 22 to sit out there and pick 'em off as they appear. Most practical, however, seems to be to avoid planting gopher food where gophers have already made their home.

So... everyone gets a very lovely tater sampler this week. They are colorful, delicious tubers. I'm sure they'll be enjoyable. There's just not as many of them as expected.

That's the news for now. Time to go work on stuff.

See you soon!

Your Farmer,

Kollibri

10.04.2007: "A Word on 'Organic'"

10/04/2007 11:13:55 AM PST
Subject: "A Word on 'Organic'"

On Monday I got some [x-brand] Cheese from the Co-op. [x-brand] is a local dairy, and is certified USDA Organic. On this particular occasion, it gave me Food Poisoning! Everything emptied out of my body on Tuesday and I couldn't eat anything at all that whole day. By Wednesday it had passed, but it took all day to get my strength back up, eating again. Today, Thursday, I feel back to normal.

[x-brand] dairy products are sketchy. I know someone who's an Inspector for Oregon Tilth. All food that is "certified organic" is certified by a certifying agency, such as Oregon Tilth, although there are many others. The agency sends out inspectors who check things out, ask questions, etc. This inspector that I know told me that he went to [x-brand] milk and that after having seen what he saw, he doesn't want to buy their dairy products anymore. Kind of a slipshod dairy, I guess, but they rec'd their Organic Certification from another agency anyway. The organic certification did not guarantee the safety of their products, as I found out. The certifying agency is paid by the farm/dairy/ranch to inspect and certify, so being a certifying agency can be a way to make money.

The "Organic" label was developed on the grassroots level by small and medium sized farmers in the 70's and 80's. Many of the leaders of the movement were around here, and are still around here, in Cascadia. I'm acquainted with a few of them. In the 90's, the big corporate food companies wanted national standards so that they could get into the organic market, which they recognized as being a growing part of the overall food market. The USDA Organic Standards were the result of this process, and many people who worked in the movement viewed the federal standards as being "watered down."

"Organic" as defined by the Feds has nothing to do with labor standards, water conservation, fossil fuel usage, or other sustainable practices. Some say the standards are being watered down further, under pressure from large farmers/packagers. I wouldn't be surprised. The way the federal law works, it is actually ILLEGAL for me to call my produce "organic" even though it not only meets but generally exceeds USDA standards.

I know many farmers who no longer bother with federal organic certification. They just choose a niche—CSA, market, local stores, restaurants—where they can build trust relationships with their customers. That's the route I've taken as well. Not only are the USDA standards not meaningful enough (to me) but certification also requires money and an amount of record-keeping that I would likely find onerous.

Anyway, getting sick from the local, raw, certified organic cheese made me think about all these issues all over again. Losing two work days will do that!

Eat well!

xoxoxo

Your Farmer,

Kollibri

2008 Season: Backyard Booty

Date: Fri, 14 Mar 2008 11:26:59 -0700 (PDT)
Subject: [sunrootgardens] Farm bulletin from Kollibri

Hello all,

This is the first email of what is intended to be a regular series throughout the 2008 growing and harvest season. You are receiving this email because you are a CSA subscriber, a Staple Crops investor, someone who is interested in helping, or some combination thereof. I appreciate everyone's participation in whatever forms it takes.

The moon was new last Friday, marking the official beginning of my Spring planting. I follow basic lunar gardening principles because it is a convenient organization method. That means that I've been working on leafy vegetables this last week. Up at The Firepit Garden, I planted three kinds of spinach, two kinds of chard, cilantro, arugula, and two mixes of mustard greens. Of the 9 varieties planted, 4 were from saved seed. The arugula, for example, came from a 2007 CSA household a few blocks from The Firepit. It was slow-to-bolt and had large leaves. I put in two 20 foot rows of it. All of these are intended for the first few CSA shares in May.

Melanie, of Backyard Bounty CSA, with whom I am farm-partnering this year, has been getting starts going, including onions and salad mix. She has the special lights for that business set up, and is also utilizing the greenhouse-in-progress at The Firepit.

Plans for the year now cover 20-some different plots. Melanie and I have added new ones with the intention of offering impressive quantity, quality, and variety in produce. Offers of garden plots poured in after a pair of articles appeared in the local press written by folks who had come and talked to me about the circumstances of my work and then put their impressions in writing. Mel and I have been mapping out the year and working out a planting schedule of what goes where when. Most major farm purchases—seeds, tools, amendments—have been made. People are stepping forward with interest in helping.

I would say that "everything is falling into place," except that I don't see it like that. I consider "order"—including the perceived fulfillment of expectation or wish—to be an illusion of the mind; not unreal, but not real enough. Where our cultural context grants us the power to change the world, I feel that I have no personal agency beyond the ability to affect my own consciousness, and that only clumsily.

So, here I am, putting seeds in the ground and we'll see what happens.

As the second-to-last paragraph shows, I was learning to be skeptical of the inventions of my own mind. All around me, Portland New Agers were insisting that "we create our own reality" merely by thinking "positive thoughts" and having the "right intentions" but I wasn't having it. A landowner pulling the rug out from under me, gophers eating my 'taters, the vagaries of weather: these were ample evidence that we are not in control and that our reality is created by multiple agencies and factors besides our own, solely. If something did happen to "go as planned," that was hardly more than coincidence as far as I was concerned, if a welcomed one. Furthermore, I realized that if I let go of taking credit for "success" then I wouldn't have to accept blame for "failure." In so doing, I could relieve myself of a lot of pressure. These were the lines of reasoning—and attempts at non-reasoning—that increasingly marked my thoughts and were applied in my actions.

Meanwhile, on the action front, Angel and I got a new partner that year: Melanie, who ran her own CSA called "Backyard Bounty." Her original name had been, "Backyard Booty" and Angel and I liked that so much better that we wouldn't call it anything else. Being a good sport, Melanie laughed along.

Melanie had started her business in 2007 on Portland's west side, across the river and over the hills. She cold-called me in early 2008 because she had moved to the east side and wanted my advice on how to find plots to garden. We got together to talk about it. After a few minutes of chit-chat, I made a spur-of-the-moment proposition that caught both of us by surprise. Since I had an abundance of plots (seventeen) and more were being offered, what if we partnered for the season and shared them? Though taken aback to receive such a significant offer from a virtual stranger, she accepted. As with Lancia, before, the idea that Melanie and I concocted was to share all the land and split the harvests between our individual operations. Unlike the Westfields debacle, however, this new scenario worked.

Partnering with Melanie was amazingly enjoyable. Besides being intelligent, energetic and fun-loving, she was open to trying out new ideas, so her participation added more fuel to the fire of experimentalism that Angel and I had already been stoking. She had different experiences and tastes and introduced us to new methods and crops. Together we accomplished things we couldn't have done on our own as individuals. I didn't have a single bad experience with Melanie and the number of good times was more than I can count.

We rode all over town together, prepping and seeding, weeding and watering, and harvesting into bike-carts. The number of plots we shared that year reached forty(!) by mid-season. It was truly a year of bounty and booty: bumper crops of fresh produce and an infusion of fresh people fed a spirit of exuberance. It felt like we were a hot knife cutting through butter.

The news that Spring made it sound like big things were around the corner in the wider world: financial meltdown, Peak Oil, Climate Change. But this sense of approaching cataclysm was tempered by a belief that the worst could be averted if there was a radical shift in priorities. Wouldn't urban farming be essential to whatever new reality emerged from the wreckage? We could be on the cutting edge of a new reality. Ours was just a little world—of gardens and bicycles and cats—but it felt like what we we were doing was potentially "important." Whether or not that was true, urban farming was the only thing we could think of doing and together we gave it our best shot.

2008 was also the year that the "urban farming" meme caught fire in Portland, Oregon. Sunroot Gardens provided a big pile of tinder for those sparks, and benefited greatly from the light and heat that the flames provided. My world changed again.

In-Depth — "Urban Farming" as Meme: The Media Blitz

On February 20, 2008, Sunroot Gardens received a gift that gave the operation a big boost and set it on a path it otherwise would not have been able to follow. It wasn't in the plan and I hadn't asked for it.

On the evening of January 30th, I held a public event in the community room of the co-op, the same one that had fired me in 2006. (Except for briefly boycotting it while they engaged me in a legal wrangle over unemployment benefits, my relationship with the place had been cordial though, as a rule, the members of the cabal avoided me when they could.) I called the event "Bicycle-Based Urban-Farming: A Primer." Was it cheeky to make a public presentation on the subject when I had so little experience? Maybe. But my learning had been quick and I was confident that I had already amassed a trove of useful information. I also believed I had an important, topical vision to share. Secondarily, I hoped that the event would help bring in more resources for Sunroot Gardens.

During the Indymedia days, I'd helped promote many events, from protests to conferences to video showings, so I knew the basics of seeking unpaid publicity. There were many free advertising venues—bulletin boards, online community calendars, telephone poles, etc.—but you always hoped someone in the mainstream media would choose to "feature" your event and give it more play. I was in luck with "Urban Farming 101"; the widely-read *Willamette Week* opted to highlight it, and even requested a color photo to include with their blurb. This bit of good fortune basically guaranteed a good turn-out.

Sure enough, on the night of January 30th, the room was packed standing-room only. I was thrilled. I gave an hour-long talk with photos projected on a big screen and then took questions. The audience was friendly, the discussion lively, and everyone seemed to enjoy it. We ended the event only because our time ran out.

Afterwards, I met two members of the press who had shown up. The first was David Ashton of the Sellwood *Bee,* a seasoned reporter of over twenty years, who asked me a few questions and set up a time to meet later. The other was an intern from the *Willamette Week* itself, a young man named John Minervini.

John was a recent Harvard grad who had moved out to Portland just a few months before. It was obvious that a bike farmer totally fit his idea of what Portland was, and he couldn't disguise his enthusiasm to be talking to me. For our interview, he brought his own bicycle and we rode around to a few of the gardens together. He insisted on seeing my quarters in Homitsu's house so he could include details about my lifestyle, which he found quite spartan. I thought it was exemplary for John, as a reporter, to immerse himself so thoroughly in his subject, but his journalistic inexperience showed in the great number of phone calls over the next few days with follow-up questions. Clearly, his editor was prompting him to get everything required for a complete story. I didn't mind.

On February 20th, the article appeared. John had done a good job of following the Willamette Week play book: "The Bike Farmer" did not sacrifice style for information, featured clever one-liners over in-depth ideas, and went for breeziness when threatened with the down-to-earth. As a representation of my work and its context, the article left a lot to be desired. As a promotion of the business, on the other hand, it was hugely successful, and that, honestly, was good enough for me at the time (though eventually I saw that the superficial treatment was not in the best interest of the urban farming movement as whole).

In the days that followed, my phone rang off the hook and my email in-box was inundated. I had not known until then that my voice-mail box could fill up and not take any more messages. In no time at all, I had more people clamoring to buy CSA shares than I could possibly provide for. I picked a maximum number of households—25—and took people on a first-come/first-served basis. I did not keep a waiting list, but there would've been over 50 people on it if I had, as inquiries continued for months. Having enough customers: source of relief #1.

Offers of land also poured in and I appreciated these just as much. Over the next few weeks, Angel and Melanie and I happily rode all over town, checking out different properties. Here we were more choosy;

some were simply too small, too shady, or too much work for too little return. Nonetheless, we accepted more plots than we turned down. Having enough land: source of relief #2.

Volunteers turned out in higher numbers, too. Now that urban farming had gotten the "cool" treatment, lots of people wanted to be involved. The number of subscribers to the farm email list grew from about three dozen to nearly two hundred eventually. Most people who said they wanted to help never showed up and of those who did, most were not actually helpful, but that's a story for a later chapter. Regardless, having more helpers was source of relief #3.

On March 3, 2008, David Ashton's story appeared in the Sellwood *Bee*. Titled, "Urban farmer rides bike, not tractor," Ashton's work was the best story that appeared about Sunroot Gardens, of all that were published. Straightforward but nuanced, it delved deeper than "The Bike Farmer." This was the difference not merely between a professional and a newbie, but also between someone with a sincere, years-long commitment to his community and someone else who was merely seeking a quick hit (but didn't know what else to do). Ashton also made easier work for himself than Minervini: the majority of the words in the article were quotations from me, and his task was mainly to string them together. This he accomplished concisely and without pretension and I appreciated that more as time went on.

The next two media hits were one-paragraph treatments in *Edible Portland* in April and the *Portland Mercury* in June. Each one was followed by an up-tick in phone calls and emails.

In July, the *Portland Monthly*—a glossy page-flipper for the liberal bourgeoisie—gave me a ring and I met with one of their writers for an interview. Later, their photographer met Angel and I in one of the gardens for a shoot. He was of the professional caliber who charges Franklins per hour and spent 45 minutes just setting up the shot—lights, reflectors, laptop—before asking me to stand in it. He posed me with a huge armful of leeks. The resulting picture was well-composed and quite flattering; he had used Photoshop to apply a glisten to my muddy arms— just a touch of glam. Angel and I found it uproariously funny.

The accompanying article was just one long paragraph, and told the story of the Westfields debacle. It was part two of a six-part feature entitled, "Eco-Freaks," a turn-of-phrase that was cute but trivializing, if not outright demeaning. My friends at Terwilliger Community Farm were also subjects. Each of the "eco-freaks" were asked to provide a quotation to include and I submitted a defense of the Dandelion with a call to cease the war on it. Their response was tepid and they asked me if I could say something else. I refused and they printed it. Though my pitch for the maligned *Taraxacum officinalis* was totally soft-pedaled, it was indeed a jab at the values of the landed gentry who bought their magazine, and they were conscious of that on some level. Regardless, my voice-mail and in-box received fresh inquires.

The final media hit of 2008 was a piece in the *Oregonian* on October 2nd. Nancy Burke penned a short blurb entitled, "Separate the wheat from the traffic," which described me threshing and winnowing wheat at the Firepit Garden and included a color photograph. It was far too brief to explain anything well, but it was one more ping that helped build a name and reputation and I welcomed that.

In January, 2009, Sunroot Gardens was written up by Marci Krass in Oregon Tilth's periodical, *In Good Tilth*, as part of a longer article entitled, "Green City Streets and Blooming Roofs," which also discussed other projects in town. It was a solid article that took urban farming seriously. This was the kind of journalism that I believed was needed and I was happy to see it.

In February, 2009, I was hanging out in the Fort at the Firepit with River, a long-time friend-of-the-farm, when my phone rang. I picked up the call and put it on speaker. It was a DJ from KBOO, Portland's community radio station and she wanted to know if I was willing to be interviewed about urban farming for the show, "Air Cascadia." I was totally psyched about the invitation; during my Indymedia days I had gained a true appreciation for KBOO and their dedicated volunteers.

So I accepted and made arrangements to come into the studio. The woman was pleasantly surprised that I would show up in person; most people did interviews by phone. When I hung up, River said, "She sounded really cute. Wish I was going in for the interview!" He was clearly smitten.

"Well then you should," I told him. "You can be 'Farmer K' for the day. Who'll know?"

He protested for a minute, then gave in. His desire to meet the face behind the voice overcame his sense of propriety. So on February 17, 2009, three days short of a year after the *Willamette Week* article, Sunroot Gardens got its seventh media hit.

I listened to the interview and howled with laughter. Not that River was being trying to funny—I just found it hilarious that the prank was being pulled off successfully. As for representing Sunroot Gardens and its mission and vision, he did a fine job—as well as I would have, as far as I was concerned. My absence was no loss to the cause and was a gain for being so entertaining.

River came around to the Firepit afterwards and told me that we had almost gotten caught. One of the volunteers at the station that day was a woman who knew both of us well and he needed to duck from her a couple times. Both of us kept the prank our own secret, and no one ever called us on it. This telling is the first public confession.

In the summer of 2009, *In Good Tilth* featured Sunroot again, this time with a story called, "Code Oranges!" that examined urban farming in the context of "food security" (which relates to the intersection of nutrition and food availability with economics and the environment). The writer, Joel Preston Smith, was a disgruntled fellow who was transparently put off by my rough manners. He managed to misspell my first, middle *and* last names and got several facts wrong (claiming, for example, that I had five employees). Regardless, the piece hit the target in its overall exploration of the topics at hand. More calls and emails also resulted.

In early 2010, a farm helper put together an excellent article about the Staple Crops Project for the nationally-distributed *Permaculture Activist* magazine. This greatly pleased me since it gave me the opportunity to deliver some pointed remarks about Permaculture directly to the Permacultists themselves in their own venue. I took on their sacred idol, "design," and blew holes in it. If I was more pleased with this article than any other it was because the farm helper wanted assistance with it and I ended up ghost-writing the majority of it for him. That it went out under his byline alone was no big deal to me (and was fun for being another prank). Getting the message out was my paramount interest.

When I had been involved in political activism, a perennial complaint had been the lack of resources. Scrambling for cash and supplies took significant energy and time. But with Sunroot Gardens, I no longer had this complaint. The media attention gifted us with what we needed materially, leaving us free to focus on creativity and innovation. This factor, more than any other (notwithstanding our motivating drive, intellectual agility and intuitive acuity), was responsible for the success that we enjoyed from 2008 to 2010.

But it was just the luck of the draw. I could have spent years seeking this kind of press and not getting it. "Timing is everything," a girlfriend of mine used to say, and this was a case where that was true. That moment—2008—just happened to be the timing when "urban farming" could become a meme and be elevated to "cool." I fully realized how lucky I was and ran with it.

04.08.2008: Cultivating a moonscape (The Staple Crops Project begins)

It always irked me (and still does) to hear a home gardener say that they "grow all their own food" in the summertime. In nearly all cases, this claim was (and is) patently false. In the USA, the typical diet is comprised of 10-15% vegetables and fruits and 85-90% grains, legumes and oilseeds (as well as meat and dairy, for those who indulge). So, at best, the boastful gardener is producing "all" of 10-15% of their food. Maybe 20-25% if they are eating that many more vegetables at that time of year. But they are not growing their own pasta, bread, peanut butter, etc., etc.

Total local food independence in actuality was the central agricultural purpose of Sunroot Gardens. Our favorite dinner was quinoa, tempeh and greens—what one friend called "the Portland Meal"—so we aimed to grow all the ingredients for it. Besides being delicious, the Portland Meal was an intelligent choice because it provides enough protein, carbohydrates, vitamins and minerals to form the basis of a nutritious diet, and is vegan and gluten-free to boot.

Hence, the "Staple Crops Project" was born, with its own budget and set of investors (of which Angel was perennially the biggest). Knowing that labor would be as important as money for the project to succeed, I created a distribution scheme in which 40% of the harvest would be divided among the financial investors and 40% among work-traders (in proportion to hours worked). The remaining 20% would be retained by Sunroot to feed the farmers, provide seed for future seasons, or to sell.

Compared to vegetables, staple crops take up a lot of space. Plots are measured in terms of acres, not square feet. So in 2008 we acquired the use of two large chunks of real estate for our experiments: "Carver," two acres we leased in the country on the Clackamas River near Oregon City, and "Hampton," one and a half acres in Milwaukie, a low-density town immediately south of Portland.

We planned to grow quinoa, soup beans and soup peas, flour corn and oilseed sunflowers (for pressing our own cooking oil). Additionally, wheat was already growing at Carver, planted by the previous farmer tenant as an overwintering cover-crop, so we planned to harvest that too. We entered the season with high hopes and excitement.

We had no idea what we were getting into.

Date: Tue, 8 Apr 2008 14:36:22 -0700 (PDT)
Subject: [pdx-urban-csa] Farm Bulletin

FARM NEWS: Ground prep and planting started for the Staple Crops project, during two different visits to the acreage in Carver. Using an escalating set of techniques (first hoes, then a 5hp tiller, then a 13hp tiller), we attempted to get the peas in the ground, which we eventually did. The land at Carver has been conventionally farmed from time out of mind, and the condition of the soil shows it. Over the quarter acre we were working, we did not find ONE worm. Not one. The tilth of the soil was okay—silty, not too much clay, if a little rocky in places—but it was clearly a lifeless medium. I believe the three varieties of peas we planted can make their way there, being the tough plants that they are, and we gave them a microbial inoculant to encourage nitrogen fixation (which is a standard organic farming practice) and vigorous growth.

Most people who visited Carver and helped out found the condition of the soil remarkable for its sorry shape. Parts of it were like a moonscape. I myself thought of the gravel lots I've seen along Vancouver/Williams in N/NE PDX, where trucks have been parking for years, and had the thought that the soil in such spots seems healthier. At least dandelions, dock, St. Johnswort, and other 'weeds' are growing there. We don't have a soil test back from Carver, but it is unlikely to show any chemicals, as most agricultural applications break down quickly. That they break down into the local water table is another issue, but the soil doesn't hold on to most of them for long. I also find it worth noting that the chemicals used in country farming are much more poisonous than anything that's allowed for use in the cities. The idea that cities are dirty and the country pristine is a delusion easily punctured through direct experience.

05.17.2008: "Orgiastic growth-spurting bliss" (and Quinoa)

Date: Sat, 17 May 2008 17:19:56 -0700 (PDT)
Subject: [pdx-urban-csa] May Full Moon farm update

Just heard some news from the Rock Creek Campus of Portland Community College, where the fellow who's been running the greenhouse there for three decades or so said that—up until this recent heat spell—there had been only FOUR days of full sun all day long since the end of February. His plants there are far enough behind—from that lack of light—that he had to put off their annual plant sale by two weeks.

That's part of the context of this Spring, which has been cool and dry and cloudy. Veggies want warmth and wet and sun. The PCC greenhouse was getting warmth and wet, but not enough sun. The outside veggies have been lacking in all three. I estimate that most of the crops planted so far this year are between three and four weeks behind. Radishes are about ready to pick, but there ain't much else, so we are hoping they will hold in the ground for a little while. Some mustard greens and arugula are pumping out the leafiness really well too. In a couple weeks we hope to get to-gether salad greens and radishes for all the CSA folks. End of May or Beginning of June, depending on how it looks.

This crazy heat of the last three days has sent some plants into an orgiastic growth-spurting bliss. Others—younger and more tender —have shrunk, wilted, or simply not come up at all. Keeping them all watered has been an impossible-to-achieve task, as many things have wanted moisture more than twice a day, due to the suddenness of it all. Easing into a heat is easier for plants, but they haven't gotten that.

The greenhouse thermometer, which is in the shade in the back of the structure, hit 110 degrees today. The tomato, pepper, and eggplant starts LOVE IT, as well as the marigolds, castor beans, and other semi-tropicals in there. I've been sleeping outside to get out of the house's stuffiness, and have been waking up 5:15ish, refreshed and ready-to-go.

The rain of a few days ago brought up the buckwheat cover-crop in three newly tilled beds: The Diamond Garden, on Harrison at 33rd; Staehli's, on 62nd near Stark; and Mall56, on Mall Street & 56th. We sowed this fast-growing plant to suppress grass-regrowth in the newly tilled beds, until it's time for the main season crops there which will be, respectively, salad greens, snap beans, and watermelons.

STAPLE CROP PROJECT UPDATE

The quinoa is up!! Yup, that heavy precipitation of a few days back, with the sun right afterwards, brought them up. They were showing no moisture-stress or heat-stress on Friday, despite the conditions. A very drought-tolerant plant, even when small! It's closely related to Lamb's Quarters, a "weed" that does just fine in these parts without being weeded, thinned, or watered. Only gently cultivated, quinoa is. It retains the hardiness of a wild plant.

The peas are up and three inches tall down at Carver, growing out a cement-like soil that shows the destructive results of decades of over-tilling with huge machinery. We'll see how it does. If anything can struggle through such circumstances and still thrive, it is field peas.

The Project still needs more land. The first two locations—Carver and then Hampton—have served as the homes for the peas and the quinoa, but are unlikely to give us anything else. Leads are being investigated all over town, including in Outer NE. At least an acre is needed within two weeks. Call/email with any ideas.

As it turned out, we did not find any more plots for the Staple Project that season, but the two we had—Carver and Hampton—turned out to be more than enough work on their own.

05.23.2008: Speech Delivered to the Village Building Convergence

In May 2008, I was invited to deliver a speech to the opening night of the Village Building Convergence (VBC). The VBC is an annual event in Portland that purports to showcase natural building and Permaculture and that attracts visitors from all over the country and the world. At the time, my annoyance with the word, "Permaculture," was growing. I was also turned off by the self-congratulatory tone of the VBC and the Permies in general, so I wrote a speech to deliver, but only literally. As a prank on the evening of the event, I brought along Troy, a regular helper who wanted to go to the VBC "to meet girls." I convinced him to read my speech in my place with the idea that he could better achieve that goal. At the time, the media-induced hullabaloo over Sunroot Gardens and 'Farmer K' was at a fever pitch, so—I assured him—he could ride that mojo like it was his own and cash in himself since I wasn't interested.

At the event, the woman at the door was hesitant to let in speakers without paying admission but we pushed past and made our way to the stage. When my turn arrived at the microphone, I stated that I had made good on my promise to "deliver" a speech but that Troy would be reading it for me, since I had beans to plant. Then I left.

I had published the text of the speech on the Portland Indymedia website earlier that day:

VBC Speech: "Key to Growing Food is Awareness"
author: Farmer K
23.May.2008 12:19

I have never attended a Village Building Convergence event that you have to pay admission for. I have never had the resources to do so. This year I have been invited to speak at the opening night, on Friday, May 23rd, which is just such an event—pay or volunteer to enter. So, to open the experience up, I am writing the speech ahead of time in order to post it here to indymedia first. Here it is:

- - - - -

I have been asked to speak here tonight because of the way I have been spending my time lately, which is growing vegetables, fruit, and herbs in a bunch of different plots around Southeast, and doing most of the traveling, hauling, etc., by bike. "Bicycle-based urban agriculture," it has been called. I am running the operation as a CSA. A CSA—which stands for "Community Supported Agriculture"—is a business arrangement in which a set of households provide resources, fiscal and otherwise, to a farmer in the Winter and Spring and in return receive produce throughout the Summer and into the Autumn.

Together with my farming partners, who also ride their bikes everywhere, we are growing food for 40 households out of all these plots. We also have "The Staple Crops Project," with its own set of supporters, which is intended to raise survival foods such as quinoa, soybeans, sunflowers-for-oil, soup peas, lentils, and more. The overall goal of these projects is food independence, for a small number of people anyway, by this winter. Also, we will share whatever it is we learn with whomever wants to know. When it comes to food growing, none of us can afford to make any trade secrets.

Many people come up to me and say they think that what I'm doing is "cool." I think thats a ridiculous thing to say for two reasons: First, the very act of deciding what's "cool" or "not cool" is the luxury of decadent society that is flabby, sloppy, ungrounded, distracted by the superficial and doesn't know how to take care of itself. I'm not interested in any of that. Secondly, all I am doing is working with the resources that happen to be in front of me at the moment. That's all that everyone is doing, no matter what their perceived position in life is, or what resources are available to them. I, personally, am not special, and certainly not "cool." I am no different than anyone else. The bike, the many gardens, the so-called "sustainable" methodology. None of it means anything. We are all the same.

I have been told that the theme for this evening is "Big Vision." I don't know how much of that I have. I'm not sure if I even know what the term means. What I will say is that, as far as I can tell, the key to effective food growing is the same as the key to anything, and that is: Awareness. Paying attention. Being present with your situation. Listening to your senses. Going with the flow.

All of these phrases refer to the same unwordable state of consciousness. It is something that cannot be learned. It can only be discovered. No one can teach it to you. Only you can find it. It is not intellectual, rational, or logical. It emerges only when the mind is clear, and not filling itself with a bunch of big ideas. Awareness cannot be modeled. Awareness is not a system. Awareness is free of dogma. There's nothing to map and no design course you can take to achieve it. Everyone is on their own with this one, no matter who they are, where they live, or what they are doing. Again, we are all the same.

Farming is mostly about the logistics of particular tasks—such as preparing beds, sowing seeds, harvesting crops, or—common here in Cascadia: clearing blackberries or responding to slugs. These tasks are often quite simple. They come and go quickly and it's on to the next the thing. In Buddhist circles, this type of activity has been described as "carrying water and chopping wood." It's the stuff you have to do to get by, and it generally involves no more than moving something from one place to another, or making changes to its perceived physical state. You cart, bury, cut, wash, or cook something, for example. It is on such logistics that the survival of the physical body generally depends. How we eat, clothe and shelter ourselves.

These logistics—how we "carry water and chop wood"—vary from situation to situation. The way I choose what method to use is by trying to find the Path of Least Resistance in every situation. In the case of the multiple garden plots that I tend, I have found that each has its own way. Soil, sunlight, and water vary from place to place. No one-size-fits-all model would work on all of them. What is entirely inappropriate in one place might be the best thing to do someplace else. I have no place for the present progressive tense in describing the work I do: that is, there is no methodological answer to the question of how I "do" anything, as in, "How do you find your plots," or "How do you plant your beans," or "How do you water?" Each situation has been its own, unique and unrepeatable. The only consistency is that I attempt to find the Path of

Least Resistance by cultivating as much Awareness in myself as I can.

Over the ages, people have suggested many different methods for cultivating Awareness: meditation, yoga, diet, chi gong, music, dance, tantric sex, psychedelic drugs, and more, including gardening. While all of these things can certainly have some effect on one's consciousness, none of them "give you" Awareness. There is nothing to receive. You already have it. We all do. It's simply a matter of letting go into it, and living from there. No one method is guaranteed, required, or predictable.

"Dogma" is nothing but the elevation of one given method above the others, despite the fact that one given method will never work for everyone. Dogma creates clutter in the mind, and crowds out Awareness. Dogma interferes with finding the Path of Least Resistance in our carrying of water and chopping of wood.

It is the blind following of dogma, for example, when, in the name of "Permaculture," one elevates sheet mulching as a preferred method. Under the standard of "sustainability," people have buried acres of soil under cardboard, mulch, and god knows what else, and named the whole thing after the lunch special at some Italian restaurant. I have never personally sheet mulched a garden plot because it has never been the Path of Least Resistance to collect and convey all those materials, especially by bicycle. For those with a single small plot or money to spend, it is perhaps not as logistically challenging. Those people should go for it if that's what they want to do. I do not recommend or not recommend any one method over another. I have noticed, though, that un-sheet-mulched plots provide a better opportunity to experience the volunteer plants that thrive in the area, and which show something about the spot. Here in Cascadia, many of these plants, both native and naturalized—and even "invasive"—are desirable or useful as food or medicine. I have a lush volunteer chamomile patch I can show you sometime as just one example.

In this way, I say that "dogma" is a kind of sheet-mulching of the mind, which buries the volunteering sprouts of our own Awareness, which are the vibrant sources of our own essential resourcefulness.

If you want to know basic gardening information, such as planting dates, seed depth, water requirements, harvesting methods, etc., you can look them up in a book, call the extension service, talk to a farmer, or use any number of other reference tools. That stuff is no big deal. Just make sure you're looking at accurate information for your bioregion. The heavy mulch that works so well in New England, the Midwest, or the South, for example, just creates slug heaven in the Spring here, in my experience. So, start with the experiences of others if you would like; then find your Awareness and you will see what works and doesn't work for you in the particular place you are in. Such clarity is indescribable; the truth cannot be spoken. All these words I have spouted tonight on the topic are meaningless bullshit.

One final thought:

These situations—of finding the Path of Least Resistance through Awareness—give us the opportunity to have an Ego-Free experience. This is because there is no reason to take chopping wood or carrying water personally. The fact that one individual knows more about one method or tool than someone else doesn't matter. The knowledge of logistics is a big pool and someone has this bucketful and someone else has that one. Working together or by ourselves, we can set ourselves free of our Ego Attachments—that is to say, of our expectations, fears, and hopes about ourselves and the world—and focus on simply getting the task done, whether that's growing food, making clothing, or building shelter. Along the Path of Least Resistance, Ego is a form of friction that blocks Awareness. It befuddles us from following the flow. Don't think about it. That doesn't work.

An acquaintance of mind has this quotation at the bottom of his email, attributed to John Barlow and Robert Weir, 1982: "The future's here, we're it, we're on our own."

Indeed. Here in the future, we can all let go—of artifice, attachment, and ego—and live from the perfect and eternal Awareness that we all already have within. There is nothing to wait for. Enjoy.

Troy said the audience received the speech with stunned silence. I was highly amused. Never mind that the message was really nothing more than Permaculture's First Principle: "Observe and interact." Such a reaction was no surprise; "Permaculture" in Portland was (and is) more of a "scene" than an actual agricultural practice or legitimate social movement, despite how it presents itself. Which is too bad, honestly, because the world really does need different agricultural practices and new social movements.

In retrospect, I would say that "awareness" was not the best word choice in this speech. What I was actually talking about was "attention," which is not nearly as challenging. Not that this edit would have changed the message's reception at the time.

06.15.2008: "Big Push Time"

This email is typical of the type that I put out at least 2-3 times a month during the summer of 2008. It is primarily a call-for-helpers, but also attempts to explain agricultural cycles and to describe the bigger picture of the current social context. In my own life, I have rarely missed the forest for the trees, but I saw then that most people do and must be frequently reminded that a wider world exists beyond the narrow scope of their daily routines and travails.

Date: Sun, 15 Jun 2008 00:30:53 -0700 (PDT)
Subject: [pdx-urban-csa] "All hands on deck" for 6-day staple-planting spree

Friday, the 20th, is the Summer Solstice. Day length reaches its longest at this point. Within a couple weeks, it begins decreasing, on its long slow slide to the Winter Solstice, when the night is the longest of the year.

It also marks the end of seeding season for summer growing crops such as corn, squash, beans, and sunflowers. These include not just sweet corn, zukes/cukes, and snap beans, but also field corn for drying, winter squash for storing, beans for soup, and seeds for cooking oil. Not mere vegetables, that is, but staple crops as well. The stuff you need to get through the winter.

We are all used to not thinking about food that way. As in, trying to grow what we might need. For decades in this part of the world, food is not hard to come by, even if you are dead-broke. It's not like in the less industrialized world, where actual famines kill many people. We've been sheltered from that, as happens when you live in a rising empire.

The Empire run nominally out of Washington DC is no longer rising, and the food situation is changing. So is the climate, and that's also throwing off a lot. Add to the global food crisis the extreme weather events of the Midwest, and the crop failures that is bringing, and suddenly it is clear how precarious our position is.

Spending the next six days putting as many seeds into the ground as possible is what I will be doing. During this period, the amount of work to be done would welcome an unlimited number of helpers and volunteers. There are at least four (and possibly five) large plots to prep and seed. The schedule will come together as it is happening and you can call me at any time to find out when/where things are going on.

So far, scheduled: Sunday: SE 30th & Grant, just north of Division. clearing and seeding large overgrown backyard. Monday: SE 81st & Claiborne, just south of Duke. bed prepping and seeding of squash and beans. Tues-Fri: Divided between those two, plus two new plots, one at 117th & Division, and one in the West Hills, just past the Zoo stop.

This is Big Push Time. It is also just a part of the agricultural year, which has a different set of priorities and cycles than the socio-economic world. With agriculture, the schedule is set for us by the plants and the weather. We must follow it. In that way, it is more real than the socio-economic world, which is a mere creation of the mind. Yet we have come to take the superficial as more actual than the deep. Not an effective way to feed yourself.

So, come out for a half hour or five hours or all six days. We will provide tools, as much food as we can, and room/board for those who enjoy immersion. Just call and let me know.

In-Depth — Havens from The Grid: All About the Gardens

The heart of our farming was in the gardens. That's where our crops were growing, of course, and where we spent most of our time, but there was more to it than that. The acts of farming—taking out a lawn, turning the soil, sowing some seeds, tending with care and finally harvesting—were all *transformative* acts in more than one sense.

On the most obvious level, the appearance and function of a space was transformed: An ornamental monoculture, a lawn, was replaced with a productive polyculture, a garden. The useless became useful. This transformation offered the immediate satisfaction of accomplishment —"a job well done"—and put us one step closer to the goal of providing for ourselves and other people.

On a cultural level, replacing grass with vegetables transformed the *message* sent by the property to neighbors and passers-by. Historically, lawns were popularized by the British landed gentry who installed them to showcase their wealth; only rich people could afford to take a field out of agricultural production and devote it to decorative purposes. These ostentatious expanses literally and figuratively distanced their idle owners from the toiling masses, from whom the real estate had been stolen in the first place. Previously, the "Commons" had been available to peasants for farming, grazing and forestry, but the Enclosure Acts seized and privatized these lands. In the new hierarchy, the lawn was a bright green line demarcating the haves from the have-nots. In our own day and age, few people know these origins, but the social impression projected by the contemporary lawn is fundamentally unchanged: prosperity, leisure, order. Conversely, home gardens can connote the opposite: poverty, labor, messiness. We were well aware of this transformation in message when we took out a lawn for a garden and were delighted to be the agents responsible. It appealed to our appetite for social mischief, like coloring outside the lines on purpose.

On a deeper level, any act of transformation affects not only the transformed but also the transformer. Movement in the material world is accompanied by corresponding movement in the mental world, a place that is no less real for lacking physical substance. By making the domesticated more feral, we were inciting *ourselves* to break with convention, to "tune in, turn on, and drop out," as Timothy Leary famously put it. Or alliteratively: Liberating land from lawn loosened our leashes. This would lead me, personally ("spiritually"), to places unforeseen.

De-lawning a yard was also a "radical" transformation in a literal sense. The word, "radical," originates from the Latin word, "radix," which means "root." "Radix" is also the etymological root word (ha!) of "radish," hence bringing us full circle since grass roots were being replaced in part by radish roots. We joked that we were an "anti-grassroots organization."

With so many gardens (thirty by 2008), each one needed its own unique designation. Whenever possible we would name a garden after a cat. "After all," I told people, "the cats spend more time in the gardens than anyone else." The particular honored feline might be the official resident on the property or a neighbor who considered it their territory regardless of the human-set legal boundaries. This policy resulted in memorable garden names: "Tic-Tac-Toe," "Ninja," "Soleil," "Prosperous Calico," "Subcomandante Ramona" and "Roly-Poly Sneezestress." We carried Catnip with us everywhere so we would always be prepared to make an offering to the guardians.

The gardens varied in size from less than a hundred square feet to some fraction of an acre: a quarter, a third, a half. Some sites we worked for just part of a single season with quick annual vegetables and others hosted perennials for multiple years. Each garden had its own unique combination of traits when it came to soil type, amount of sun, availability of irrigation, etc. No "one-size-fits-all" approach could be applied across the board for designing or caring for them, much to the consternation of people who wanted a quick answer to the question, "How do you do such-and-such?." Creativity and flexibility were essential and I greatly enjoyed the challenge and variety.

Growing food for a business lent a serious, often urgent, tone to our work in the gardens. After all, people had paid us up front; we wanted them to get their money's worth and not have anything to complain about. (We soon learned that some customers would always have complaints no matter what we did, but they weeded themselves out over time.) With production as a prime motivator, the aesthetic appearance of our plots took a back seat. This led to dissatisfaction for some landlenders (as is discussed in "The Versailles Syndrome: All About Landowners") but with CSA shares to fill—and the task made logistically complex by farming in a "distributed network" of plots—there was only so much we could do, and we generally deferred to the needs of the business over other concerns.

We were also fascinated by the possibilities of agriculture as restorative stewardship. That is, how could farming help remediate the disturbances of human activity and recreate more ecologically balanced spaces, ones with healthy, living soils, and a population of complementary plants that provided habitat for non-human creatures? And how could such spaces be set up to require as little human interaction as possible, or even be entirely self-perpetuating? And last, but certainly not least, how could such spaces reliably produce food, season after season?

Many people believed (and still do) that the answers to these questions are found in what is called, "Permaculture." But after investigating, I found that school of thought lacking. By its originators, Permaculture was presented as a scientific approach based on design. I found the intellectualism cold and the emphasis on design hubristic, to say nothing of its modus operandi of cultural appropriation. In Portland, the "Permies," as they called themselves, had injected "earth-based" religiosity into the mix, making it even more unpalatable for me. As someone who was (and is) deeply spiritual (okay, I admit it), I found (and still find) New Age beliefs and practices offensive for their vacuity, superficiality, egotism, clique-ishness, and unexamined First World entitlement. That Permaculture could so easily absorb such bullshit revealed to me its lack of center and substance.

Our approach to agriculture-as-restoration, by contrast, was what I would now call "re-wilding." I had not heard the term in those days, or undoubtedly I would have used it. We applied a light hand to a space (after completely removing the lawn grass), broke the rules of conventional gardening freely and encouraged the return of undomesticated plants. We also closely observed what happened and compared results from spot to spot. Common practices include leaving patches of annual greens to seed

56

and resow on their own schedule, installing hardy perennial medicinal plants, and eschewing autumn "clean-ups" in order to provide winter food and shelter for birds.

When we encouraged the return of undomesticated plants, what were doing was welcoming "weeds" alongside our crops. Permacultists often smothered the soil with sheet-mulching, but I was always curious about what the native seed bank contained. I knew that "pioneer species"— those plants that thrive in disturbed areas as the first wave of reclamation —often played rejuvenating roles by providing ground cover, restoring nutrients, and making habitat. Plus, many of the pioneers were edible or medicinally useful for humans: Cleavers, Clover, Cress, Miner's Lettuce, Mullein, the Plantains, Self-Heal, Wild Lettuce and our favorite, the Dandelion.

Ah, the maligned *Taraxacum officinalis*! It provides nutritious greens in the late winter, medicinal roots in the fall, and flowers in the spring that make a delightful white wine. Bees love the blossoms, the taproots break up compacted soil and the leaves mulch the ground. An entirely useful—and beautiful—plant that deserves none of the malignity dumped on it. Indeed, the hatred that people in the USA hold for the Dandelion is nothing but an unctuous attempt to deny their own violent, collective domestication.

We actively cultivated Dandelions, leaving them wherever we found them except when we transplanted them into rows for easier flower harvest. This appreciation and these practices did not endear us to many landlenders, and more than one relationship was hence revealed as not a good match. In that way, the Dandelion provided a quick and easy barometer for social relations that only deepened our bond with it.

As time went on, the role of the gardens as havens from city life became paramount for me. By immersing myself in these vibrant spaces, many of which were half-wild by the end, I found a soothing balm from the noise (both aural and energetic) of the city. I flitted from one to another, spending as little time as possible in the places in between.

In the end, it was at the personal ("spiritual") level that I most strongly appreciated the transformative nature of the gardens:

Date: Fri, 30 Jul 2010 22:48:02 -0700

...I have certainly had my deepest moments of personal joy in the gardens I have stewarded around the city. Life With The Plants can inspire a focus that puts the so-called "Real World" of rules and limitations into perspective, revealing it for the farce that it is. "The cycles and priorities of farming are completely at odds with the cycles and priorities of urban life." I have said that many times. It was with the gardens that I personally found Haven in this crowded, polluted cage we call a "city." Indeed, the beauty of the interplay between self and nature is vividly illustrated in these places, and they are just about the only places I spend time in, in the city, anymore. At this time of year, I sleep in them too, surrounded in dream-time by the energies of other creatures, plant, animal, and otherwise. My own sense of "Self," in the ego sense, has been dissolving into this interplay, and with it many worries about the world. Laboring directly for one's means of survival offers the opportunity to skip the funny business—but not the monkey business!—with a release that can feel like a full body sigh of contentment.

06.24.2008: "The Ultra-Magnetic Stoplight Crop Circle"

A tall-tale email, but totally a Permie/Burner wet dream.

Date: Tue, 24 Jun 2008 00:22:34 -0700 (PDT)
Subject: [pdx-urban-csa] Planting of Fall/Winter crops begins

The Solstice passed a few days ago and a shift is underway. It is as if a giant wheel has slowed to a stop and is now beginning to reverse its direction. These long days give plenty of time to work and there is plenty of work to do. Parsnips are being seeded this week, for harvest over the winter. We're also putting out summer squash and cucumber starts, tiny still but ready to burst once they have their feet in warm soil.

A "Three Sisters" garden was seeded at the Waverly plot, which is in Milwaukie, just past The Bins. "Three Sisters" is a planting of corn, squash and beans together, and is a traditional indigenous agricultural style, predating European invasion. The beans provide nitrogen to the corn, the corn provides something for the beans to climb, and the squash runs along the ground, suppressing weeds and helping to hold down moisture. There are many variants on the style, and we used one suggested by Tom, a helper, that he learned while visiting Peru last year.

It seems that Three Sisters gardens there were laid out based on sacred geometry, in order to attract and focus certain energies for enhancing plant growth. Also, the colors of the vegetables were chosen carefully, with their symbolism reflected in the name given the planting. The patterns of smaller plots fit into greater patterns of the surrounding plots, making designs that would have been fully appreciated from the air. This entire tradition survived the

Genocide of the conquistadors, albeit on a smaller scale, and began to gain attention again in the 1950's. It is practiced today throughout Peru, even in the barrios of Lima, and hence the connection to Tom.

It seems Tom met a Peruvian fellow who we will call 'Bolivar' at Burning Man a couple-three years back, and they and their crews hung out, painting their bodies, dancing all night, setting fire to what-have-you, etc., such as happens out at Black Rock every summer. They stayed in contact via email afterwards, so when Tom found that a trip to Peru was in his future, he made arrangements to see him again.

It seems Bolivar was at the center of a scene there where nights were spent at raves, but days (well, late afternoons anyway), were spent in the gardens of the Lima barrios, where Bolivar had grown up, being part indigenous himself. Some kind of synthesis between indigenous culture and the rave scene was percolating in the narrow alley plots, the rooftop beds, and the cliff-side terraces. They all had names, most of which don't translate well into English, but which might sound like "mega-electric flower mountain" or "deep-earth shadowy ripple-pond" or "waterous-joy in the rainbow traffic island." And Bolivar's own particular garden—on the slightly slanted top of an abandoned coal shed—hosted a layout called the "Ultra-magnetic stoplight crop circle," so named because the shapes were circles arranged in threes like on a stoplight and the colors of the corn, beans, and squash, were green, red and yellow.

So it was this design that we mimicked at Waverly, using Oaxacan green dent corn, red Hidatsu beans, and golden acorn squash. It's quite a thing to see. Each "light" is made up of three concentric circles, connected with spokes. From the first to the second ring there are four, for the four cardinal directions. From the second to the third there are twelve, for the signs of the zodiac. The squash are planted at the centers and the corners, and the beans and corn are planted on the spokes and rings throughout.

Stay tuned to see where this story goes next.

The helper named, "Tom," later became known as "Tom of Mall56" and figured large in Sunroot Gardens.

07.18.2008: "Nutria Emergency!"

Date: Fri, 18 Jul 2008 13:21:35 -0700 (PDT)
Subject: [pdx-urban-csa] July Full Moon: Nutria Emergency!

Greetings, from Farmer K.

We have experienced several crop failures this year—peas from rot, beans and corn from low germination, radishes from bolting, others from weather, pestilence, accident or circumstance. And, down at the Lucky Cats! Garden in Sellwood, multiple things have been eaten by Nutria (scientific name "Myocastor coypus"), "a large, herbivorous, semi-aquatic rodent" of which more than one lives in Johnson Creek, along the edge of the 1/3 acre property. For some people they have a certain cuteness in the semi-awkwardness of their movements, the nerdiness of their faces, and the roundness of their body. "Herbivorous" means they like to eat vegetable matter, including vegetables.

At first, the little beasts were hitting just the cabbages and broccoli, and only a few at a time. Those crops were planted in small amounts to begin with, and we weren't sure if there'd be enough for the CSA folks anyway, so though it was a loss, it seemed no big deal. At night we would see "Brother Nutria" and his comrades, trundling around amongst the weeds, then out poking at the big brassicas. Personally, I figured that once they were done with the cabbage and broccoli, that they wouldn't be a problem. Angel suspected otherwise—that these were merely their favorites, and that they would go after other crops when these showy brassicas were run down. And they did.

Yesterday, Thursday, we visited Lucky Cats! to see how the water timer had been working on the parsnips beds (it had not been as the water faucet had been turned off) and we found immense new damage from Brother Nutria & family: Remaining cabbages (about half a dozen): munched; baby beet bed: razed; wild amaranth and lambsquarters: chewed to the ground; maturing chard: decimated; and—most significantly of all—the sweet potato crop of 54 plants gnawed down to stems.

Now, the sweet potatoes were ordered as "slips" from Johnny's and were planted quite some time ago in carefully prepared raised ridges. "Slips" are live root stock, not seeds, and they cost some $. Originally, over 100 plants were put in. The attrition rate due to cool weather and other lacks had hit nearly 50% already, reducing the population by about half. I had just spent half a day going through the bed, weeding it, re-mounding all the plants, and re-setting the irrigation system for them. The viney plants seemed poised to take off and deliver to us loads of sweet earthy tubers. (Angel and I had grown them once before, in 2005, and enjoyed them immensely.) But the Nutria—in just a few minutes probably—set all but 4 of the plants back, perhaps permanently.

Most spookily, we found two munched kale plants in the further-back-from-the-river part of the plot, further back than the Nutria had struck before. In that direction are tons of winter greens, about 4-6 inches high, that are 6-8ish weeks old and sizing up for over-wintering, now looking more and more like Nutria dinner. What we had presumed safe was suddenly under imminent danger.

So, here it is, nearly 4:00 in the afternoon, and we've got other things planned (such as trellising the pole beans that have three foot runners, tilling more ground for winter crops that need to be seeded now, garlic harvesting, etc., and other time sensitive tasks) but it's clear we need to do something about the Nutria. It is a bona-fide "EMERGENCY."

Fencing seemed like the best option but we didn't have enough fencing or enough poles. I stated that I was prepared to stay over at the garden that night with a tool I call "The City Hall-er," and take care of the Nutria that way, since they have so far been fearless to close approach by humans, and would hence be an easy target.

Angel suggested we build a fence, so we set about that. With over 200 feet needing to be fenced, less than two hours of open hardware store time, the closest bike cart a half hour away, and a little over four hours of daylight, we started.

Back at the Firepit, we hitched up a bike cart. On the way to the hardware store, it became clear it had a flat, so I turned back to fix it, and then meet Angel & Stella at the hardware store. I did that, brought the pump along, and biked down to the hardware store, but they were nowhere in sight. I called and they had gone to another, where they had purchased the entire stock of rabbit and chicken wire and posts and were sitting outside waiting for me to pick it up. It wasn't enough wire or posts so they asked me to get more at the other hardware store, but I didn't have the cash. So I headed up to their store with the cart while Angel headed down to mine with money. We passed each other on Harrison at 30th and slapped hands at the meeting-in-motion. Upon arrival at that store, I found that the previously flat tire now had an alarming bulge and was about to burst, due to a twist in the inner-tube. I quickly removed the wheel, deflated the tire, readjusted the tube, blew it back up and hit the road with Stella. We then stopped at the pea garden at 28th & Long and yanked the poles there, which required softening the earth around them with the hose first. We needed these too since the stores didn't have enough for us. These poles we lashed to our bike frames, since the bike cart was already overloaded.

We took off south on 28th toward Reed. The bike cart was sagging under the weight of it all and fish-tailed on curves and speed bumps. It was barely in control really. And right there, where 28th curves around east near the Rhodie garden, was a kitten in the road, that seemed to have been hit by something. We pulled over and I went out to the kitten, walking directly down the lane toward traffic, directing it away into the other lane so the kitten wouldn't get squashed more. I bent down, touched it, and made to cradle it up. Spasms went through its body, soundlessly, in what I later recognized as its moment of release. I brought it to the closest house and curled it up under a tree and left a note on the door on the back of one of my business cards, with a pencil I found lying on the ground in front of one of the hardware stores and picked up out of some intuitive flash.

So that was that for the cat, and we moved on.

On the way to the Lucky Cats! garden you can stop and view the property from the other side of Johnson Creek. On a whim, and with Stella's encouragement, we decided to stop there and "javelin" the fence posts over the creek to the garden. (Stella was the helper who showed up that afternoon for the emergency.) We tried different throwing techniques, both right and left handed, and only nine of the twenty-some landed in the river, which is shallow (less than waist deep) at this time of year, so they were easy to fish out.

As the sun disappeared below the trees we pounded in posts and started stretching fencing. Three people was the ideal number for the task and we were grateful that Stella was such a great sport, showing up into the middle of such a high-energy situation and playing it so level.

We put the posts 8-10 feet apart. We affixed the fencing to the posts with wire. We bent the bottom of the fencing outward and weighed it down with rocks (Nutria dig!). We peed at the corner to mark our territory. And we placed a bet on whether these precautions will actually work. (I've got a dollar that sez they don't.) The full moon had risen and the sunlight completely disappeared by the time we finished, a little after 11:00 pm, finding our way around the project with bike lights.

That's when a Nutria showed up, and he nosed around among the weeds right up to the fence, as fearlessly as usual. Angel told the animal about the fence. I showed the animal the City Hall'er. He ran back into the bushes when Angel stepped over the fence toward him. We'll see how it goes.

We didn't have more trouble from the Nutria at this garden because we dropped it from our network shortly thereafter. The landlender was out of town when these events happened, and though he had been talking about building a fence to protect the crops, he was displeased to come home and find that we had done it. He said we had overstepped our bounds and told us not to leave anything at his property that we valued. Our crops were "valuable" to us, so, taking him at face-value, we transplanted or harvested everything, removed the fencing, and told him we were done. This seemed to catch him by surprise but he didn't back down either. In retrospect, I don't think he meant what he said literally but I know we didn't want to deal with anyone who would throw around words that lightly. Ultimately, there were no hard feelings and we remain in friendly contact.

09.21.2008 + 09.26.2008: "Report from a DIY local wheat harvest"

Here are two articles I posted to Portland Indymedia about our work with the Staple Foods Project. "DIY" ("Do-It-Yourself") was a meme (and movement) in Portland at the time.

Report from a DIY local wheat harvest
author: Farmer K.
21.Sep.2008 21:42

On Saturday, while the news was full of stories about a collapsing financial system and possible Greater Depression on the way, some of us spent the day out in a field harvesting wheat by hand. Our methodology was ad hoc and primitive; the harvest of "spikelets"—the heads of the wheat containing the wheat berries—was substantial. Much attention has been given over to growing vegetables and scoring fruit in the city, and many creative approaches tried out; yet those foodstuffs comprise only a small percentage of the typical human diet, which is heavy on grains augmented with proteins (the majority of it vegetarian). The less-than-an-acre plot we harvested on Saturday is part of one attempt to address this issue. Help is still needed winnowing, and wheat will be shared with those who put in time.

The wheat field was about 2/3 of an acre and located outside Carver, a so-small-you'll-miss-it-if-you-blink hamlet on 224, out Estacada way. We had leased the land from its nominal owner for the year to attempt to grow some staple crops. As it turned out, the previous farmer had sown wheat for an over-winter cover-crop so when it came up, we left it. As the year went on it got taller and produced big fat spikelets. Last week, we got a call from the folks living there that the wheat was ready and did we want it. Looking at the weather, I saw that if we wanted it we needed to get it soon as rains were on the way. Sure enough, today's downpours showed that we got it with perfect timing; that is to say, just this side of too late!

I put the word out among folks who are involved with the CSA I farm for, and among the permaculture crowd round these parts, and to other people I know. Over the course of the day, two dozen people helped. We pulled the spikelets off by hand into bins, then transferred them into huge Stumptown coffee bags for transport. Adding to 13 bags worth collected the previous Tuesday on a test-run, we got nearly three dozen more!

Winnowing has so far occurred by spreading the spikelets out between tarps and stomping and "doing-the-twist" on them. This breaks the berries out of the spikelets and removes their inedible hulls. This step could stand to be improved somehow with a simple machine, or maybe by renting a steam roller! We're looking to find better ways. The next step—cleaning—has gone well. We have just been pouring the wheat/chaff mixture from bucket to bucket in front of an electric fan. (The right wind would also work for this.) The chaff blows off while the wheat berries fall straight down into the bucket.

We have taken some of these berries, ground them into flour, and made bread and pancakes with them. DEE-LISH!

Now, we've got thirty some bags of spikelets still, under plastic under a big fir tree at a house in the Hawthorne neighborhood and we are seeking more help with the winnowing. We will have berries and flour on site to take home for people if they want something then, basically just to taste or play around with. We will also be disbursing the harvest based on number-of-hours-worked, once we have a total cleaned amount and can do the math for dividing it up.

This whole thing is happening as part of "The Staple Crops Project," more details of which can be found here [link].

Local DIY Wheat processing yields over 600 lbs.
author: Farmer K
26.Sep.2008 16:37

As reported on this website a few days ago, a group of us went out and hand-harvested a bunch of wheat from a field outside town. We have since processed it, which involved threshing and winnowing. This took a few days, many people, and yielded over 600 lbs. of wheat berries! Now it is being distributed through a creative scheme that rewards monetary investors and work-traders alike.

We threshed it on tarps in the driveway under a fir tree in the Hawthorne district. We winnowed it in front of electric fans, pouring from bucket to bucket. One helper found that a basket with just the right holes in it worked well to sift off the straw and big pieces before bringing it to the fan. This discovery led to cleaner berries and less time in front of the fans.

The grand total for the harvest: 636 lb.!

Under an arrangement that was hatched this winter, 40% of that will be divided among the people who spent time helping—work-trade, 40% will go to the people who donated money to cover the monetary budget of the project (which is multi-crop, not just this wheat), and 20% remains with the farm, to feed the farmer and provide seed for the next crop.

The way it broke down: people who helped will get 1 lb. for every hour they worked. The monetary investors are getting 24 lb. for every $250 they invested (and they get shares of the other crops coming in too). The overall productivity rate—from harvest through winnowing—was 2.6 lb. per hour. 42 people helped altogether, contributing a mean average of 3.5 hours each.

All these numbers are in spreadsheet form in the attached PDF, if you'd like to see more details.

If a wheat crop is undertaken again, it would likely take less time to harvest and process, just due to shortcuts we learned this year. Also, we have seen that some mechanization with simple machines would be helpful at all stages, but especially at the harvest part; most of the crop was not brought in. Animal- and bicycle-driven methods would likely be best for projects of this scale (i.e., an acre or two). How are folks like the Amish doing it? That's about the technological style we can expect to be finding ourselves using in the coming years, what with Peak Oil, socio-economic break-down, etc.

Feel free to contact me regarding the upcoming quinoa harvesting and processing.

Sunroot Gardens FAQ

With all the press coverage bringing so much attention to Sunroot Gardens, I soon found myself answering the same questions over and over. Wearying of the repetition, I wrote this FAQ and posted it on the farm website.

Frequently Asked Questions (FAQ)
answered by Farmer K

Attention reporters/students/videographers, etc.—you are advised to read this FAQ before coming to interview/film, etc., and then bring different questions with you.

Q: So, Sunroot Gardens plants veggies in front, back & side yards. How's that work? Do the people living on these properties get produce, or what? What about the water bill?

A: Every arrangement is different. Some households want produce. Others just like that they don't have to take care of the area. When an owner or renter offers their land to Sunroot Gardens, we ask: "What do you want?" That's how we make our deal together. If the owner/renter wants something that we at Sunroot don't want to do, or vice-versa, then we don't work together. It's that simple. There's no reason for anyone to do something they don't want to do. We are all free agents.

Q: What's that mean that Sunroot Gardens is a "bike-based business"?

A: We bike around most of the time and have carts for tools and harvesting. It's cheaper, more flexible, and funner than cars/trucks. But we don't hesitate to use vehicles when they are readily available for especially large loads. It's about path-of-least-resistance, not ideology. I'd say we're over 90% bike-based.

We also want to make some new pedal-powered contraptions for heavy and/or awkward loads—if you're a creative bicycle-tinkerer and want produce, contact us.

Q: Bike-farming in the city! What a cool idea!

A: No it's not. First, just the habit of labeling things as "cool" is the luxury of a pampered class that doesn't know the first thing about living with responsibility, and I've got no interest in it. Besides, nothing is "cool" in a world where everything is perfect (and that's the world we live in).

Second, would you tell a carpenter they were cool because they used a hammer to hit nails? What else are they supposed to use? Here we are in the city, on the edge of a crumbling empire, with the conventional food system about to collapse. Where else are we supposed to grow food except all around us? Feeding ourselves is not "cool"—it is a function of being a human animal.

Third, it was never an "idea" for me. I just found myself doing it. "Life is what happens to you when you're busy making other plans," John Lennon sang. Credit for coming up with "The Idea" goes to Martin at City Garden Farms, who found it one night after 5 pints. Credit where credit's due, right? To the 5 pints!!

Q: I don't have land to offer and I want produce. How do I get some? Do you sell it somewhere?

A: You can also offer labor, goods or services to the farming efforts in exchange for produce. On a day-by-day basis, people who come out to help are offered whatever produce is available at that place and time. Additionally, starting in May/June, we will harvest produce weekly and bring it to a central location in Southeast for distribution to helpers and other friends of the farm. You will find out the details of this event when you are invited to it.

Examples of goods & services that the farm needs: bicycle repair, cart-building, heavy-load transportation (use of vehicle), plant starts, pots & trays, seeds, plastic and glass for cloches & greenhouses, compost-brigade, home-crafts, scavenged wood or tools and special supplies. If you have something to offer for food, bring it up.

Q: Yeah okay, but do you sell produce somewhere?

A: Sunroot Gardens has plans to start up a new farmers market sometime in May/June. It'll be on Sundays at a Hawthorne location. Produce will be set out there with prices clearly marked, so if you really only have US$ then you can come haggle there. (Or barter!)

We can also set up deals on a case-by-case basis. For example, Sunroot Gardens has sold coolers of produce to folks going out to the Burning Man festival. Depending on your desires and timing, we'll see what we can work out. If we know about a special event ahead of time, we can plan for it.

Really, we're open to any number of arrangements. We see no reason to limit ourselves—or you—to any specific guidelines right now. Again, we're all free agents here.

Q: I want to learn more about urban survival gardening. Can I come help? What days are you in the field? Do you have work parties or what?

A: We're workin' everyday somewhere, from now 'til the November rains, sometimes with multiple crews in different locations simulta-neously. "Work Parties" are announced for special, large projects, via the Sunroot Gardens email listserve. If you would like to be on that list, and receive emails 1-3 times per week about what's hap-pening with the farming, including when/where produce will be available, email Farmer K at kollibri (at) riseup (dot) net.

You can also just give a call about any day you'd like to join us, and we'll tell you when/where we'll be, doing what. In the Spring that'll be planting all over the place, sometimes breaking new ground, sometimes working in established places. By June, we'll be harvesting on regular days, multiple times weekly, as well as weeding, watering, planting winter crops, etc.

Q: How much land does Sunroot Gardens own?

A: Not a square inch of it. Everything is on private land, in people's yards and unbuilt lots. We do not pay rent on any of it, either, in-stead working out other arrangements—with the exception of 2 acres of country land that we are leasing for the Staple Crops Project on very reasonable terms.

The Sunroot Gardens project acts to steward or caretake all these plots around town at the present time. We assume that someday they will be watched and cultivated by other farmers, perhaps un-der totally different arrangements. The priority at this point is not to acquire anything, in the capitalist sense, but to break ground and get plants growing. The more land is under food production, the better. The conventional food supply system really is on the edge of failure, so getting localized food networks re-established is essential. Dollar says it's already too late for any of this to make a significant difference, but I guess we'll see.

Q: How'd this whole thing get started?

A: I had six or so sun years

for my first garden in this body

found an Indian arrowhead that was white.

Plant spirits have lined my path in various morphologies ever since

everywhere & when I've lived

if only as a few things on a fire escape.

Now I'm doing "this."

On my own?

or did the plants plant me here?

Q: No I mean the whole urban farming thing. With the bike. In people's yards. You know, the "cool" part.

A: Oh, that again. If you want some sort of story, read this one in the Portland Monthly or this one in the Willamette Week. Or this one on portland indymedia, from back in the growing-in-five-gallon-buckets-on-a-porch days.

I'm not going to bother to get into it myself. I don't see why it matters. It's not like there was some "method" that someone else could imitate or something. 'Twas juss life happening. Wanting to know "the story" seems like voyeurism, and that kind of thing is none of my business.

I am considering the subject closed.

Q: We heard you've grown quinoa.

A: Well the quinoa grew itself. ;-) I do recall being present for certain moments, however, such as the seed going into the ground, some weeding & thinning, and the harvest. Gregorian Calendar Year 2008 was the fourth season that quinoa and I were companions in these ways here in the City of Roses. For photos of this year's crop, in Milwaukie, see this gallery of photos on the Sunroot Gardens blog.

Q: What's quinoa?

A: An ancient South American grain, quite high in protein, in fact nearly a meat-replacement. Grown as a basic staple crop by the humans in South America until the European Invasion, when the Spanish tried to make them stop as part of their genocide campaign. Chenopodium quinoa has experienced a popular resurgence due to its ease of cultivation, immense nutrition, and wholesome nutty flavor. In the kitchen it is treated like a grain, filling a place such as rice, millet, or polenta. Being a seed, it also contains healthy oils. Each "grain" of quinoa is naturally coated with something called "saponins" that must be removed by soaking & rinsing before being cooked. These bitter-tasting compounds protect quinoa from predation by birds or insects while in the field. Quinoa is an essential part of the so-called "Portland Meal," which is why it was central to the Staple Crops Project.

Q: What's the Staple Crops Project?

A: Fruits and veggies typically make up 10-15% of a healthy agricultural human diet, with grains & proteins making up the other 85% (with exceptions). Therefore, in 2008, Sunroot Gardens took up a $ collection from half a dozen investors and set about acquiring seed, tools & equipment, and use of land. That year, we got over 600 lbs. of wheat. Other crops, such as quinoa and Canellini soup beans and flour corn, did well enough to provide seed for 2009's plantings.

Being that labor is a key to planting, cultivating, harvesting & processing field crops, a significant percentage of the final product is divided amongst the people who helped, based on how many hours they worked. The percentage going to "the farm" (for seed and for the farmers) is a smaller portion of the harvest than what the laborers and $-investors get. This is not hiring anyone—this is the sharing of labor and the fruits of that labor by free agents acting on their own. It is a purposeful break-down of the modern agricultural model in which "owners" "pay" "workers." It's over for that model, for us anyway: time to get back to the community taking care of itself.

Q: You grow medicinal herbs too?

A: Yes. The Western medical establishment is dangerous to human health. Did you know, for example, that medical errors are the leading cause of death in hospitals, killing nearly 200,000 people per year? Or that use of anti-depressants leads to violence? Or that mammograms cause breast cancer?

That's all just the tip of the iceberg, really, when it comes to what is called "modern medicine." Better to live well and find good plant friends. Everything is there in nature. So we are growing medicinal herbs wherever we can. We harvest and process these plants as we can, but often seek out experienced herbalists who appreciate the material and work with it to bring out and preserve the constituents. This is done on a "gift economy" basis, rather than barter.

Q: *What* are you smoking?

A: Raised & cured m'own smokin' tabacky, I did! One variety is a Cuban leaf, in its fifth generation in Portland, a bona-fide regionally-adapted, local tobacco! Not for sale, but tastings are offered. Will share 2009's take with those who help plant, raise, harvest & cure. Got a shed we can use for hanging bundles? That'll cut you in, too. Advice from more experienced growers also appreciated.

2009 Season: "A Call to Pitchforks!" / "50 Gardens (and counting)"

In 2009, Melanie scored a plot that was big enough to grow everything she needed for her CSA without having to run around—and she could live on-site, too—so she quit the bicycle-based farming. I understood her choice but it was sad to lose her regular company. Any sense of abandonment soon dissipated though, as 2009 brought more fresh blood to the scene.

The first person was Demeter, a lovely trans woman I took to calling, "Auntie," who moved to Portland in January, 2009, for the specific purpose of getting involved with bike farming. We initially found each other on the internet and acquainted ourselves over email. She planned to sleep in her car at first and needed a landing pad when she arrived so I invited her to park where I was staying that winter.

I still had the trailer I'd picked up in 2007 for the Westfields farm. In October, 2008 I got permission to keep it on a friend's property in outer Southeast to serve as my cold weather pad. I dubbed it, with ironic intention, "The Winter Palace." But Demeter pointed out that the Russian Revolution had been sparked at the Winter Palace, so maybe the name was more apropos than I had considered it. Regardless, it was a far more humble abode than its namesake, if cozier. I decorated it with Christmas lights, installed stereo speakers, and brought in an electric oil heater. The driveway was very long so I had privacy from the street.

Every morning Demeter would come up to the trailer for coffee and breakfast, and then for dinner at the end of the day. We ate good food, plotted mischief, and listened to lots of Miles Davis (exclusively from his "electric period," 1969-1975). Sunroot Gardens had a multitude of garden plots so I gave some to her. After my CSA filled up for the season, I passed along new inquiries to her with my personal recommendation, so she was able to fill her roster without too much work. I also had tools, seeds and other supplies in abundance to lend and gift.

I deeply enjoyed having so much wealth to share. It was truly gratifying to give things away and keep only what I needed. I perceived that my own success was due to the generosity of others so it only made sense to be generous in turn. Of course, viewed by conventional standards, I was dirt poor, legally below the poverty line, in fact. Nevertheless, I felt (and truly was) rich. Listening to the Beatles as a youth had convinced me that money can't buy what's actually important in life, and this point of view served me well. Financially, Sunroot customers (and Angel's personal contributions) brought in just enough that I was free to do what I wanted to do and then to see if happiness (or some reasonable facsimile) would follow.

Demeter was impressed at how the urban farming "Primer" at the Co-op in January 2008 had provided such a springboard for me and Sunroot Gardens, so she took the lead in setting up a similar event for 2009. With revolutionary zeal, we dubbed it "A Call to Pitchforks!" Here is the invitation I sent out to the Sunroot list:

Date: Fri, 13 Feb 2009 11:43:57 -0800
Subject: [sunrootgardens] A Call to Pitchforks! Portland Urban Farming 2009 event

Meet farmers that are changing Portland's urban landscape as they grow food in all kinds of places. Learn about their methods, philosophies & where we're going with urban farming. Q & A discussion panel with farmers, land-lenders, CSA subscribers, community organizers, and other working to create an urban foodshed in the City of Roses.

If you want to: become more independent in your food choices, get more involved with local food, and find out what's happening with urban farming, JOIN US!

A Call to Pitchforks by Farmer K

Like mushrooms after a rain, new urban farming efforts are popping up all over Portland. How appropriate in a bioregion well-known for its fungal fecundity! Yet the emerging activities are not limited to the City of Roses, nor even Cascadia. I have heard of or from other urban farmers in San Francisco, Los Angeles, Bisbee, New Orleans' 9th Ward, Philly, and Detroit. In this last city—experiencing an urban collapse decades old that foreshadows the fates of other metropolises—folks have been planting in empty lots for years. And in other parts of the world, outside the U.S., urban farming is and has been a part of the urban fabric from time out of memory. In "less developed" nations, people grow where they can, and move it around as they're able, by bicycle or rickshaw or mule. But being industrialized is no excuse for not farming: In Japan, mini-farms dot the cityscape, and broccoli and greens and melons share Main Street footage with cafes and stores and train stations. Many have already heard of the example of Havana, Cuba, where at least 50% of the vegetable and fruit produce for the city is grown within the city limits.

Here in Portland, we have a rare and delightful climate for food growing. Even though summer classics such as tomatoes, peppers, and eggplants can be tricky here (and even zukes, cukes, and snap beans in a mild summer like 2008's) it's more than made up for by the ability to pick fresh produce year-round. Our lack of hard-freezing here means we can pull up carrots in January, and see brassicas germinating in February. It also means we have more slugs, but as someone farming, I will take these slimy, slow-moving, beer-loving creatures over the grasshoppers, cutworms, and bean beetles of the Midwest any day!

I have been growing vegetables in the city only since 2004, starting with three dozen five-gallon buckets on a sunny porch. Going into 2009, the Sunroot Gardens enterprise I started has gained access to over 30 plots around Southeast, totaling just over an acre altogether, for planting fruits and veggies. This is besides the nearly 4 acres being used for growing staple crops. A two-acre chunk of this is the only plot outside the urban growth boundary, and is the sole real estate where money is being paid for a lease. The vegetable and fruit cultivation in the city is happening in backyards, side yards, front yards, "empty" lots, and along "unimproved" roads.

Sunroot Gardens has put the lie to the claim that one must own land to farm. No, you don't. If that's the only thing holding you back from investing yourself in agriculture, then you have no excuse. And here in Portland, many other folks are doing the same thing.

There's plenty of open land to go around for those interested in growing on it. Whether there is enough land to feed the people in this city is a question. Personally, I have no confidence that it is (or is not).

* * *

I am planning to be on the panel at the Portland Urban Farming 2009 event at the Co-op on Wed., Feb. 18th, to share my experiences with Sunroot Gardens. Other folks involved directly with their own agricultural efforts will also be there, and co-operation has already been occurring among us. This is a truly grassroots thing happening here. That is, grass roots are being replaced by carrot, kale, pea, pumpkin, and quinoa roots. As much as possible, as quickly as can be done, here at the edges of a crumbling empire. This scenario—of community re-localization in the space opened by the disintegration of governmental centralization—has played out countless times before with varying details. We are nothing special.

Dan & Martin up at City Garden Farms, another urban CSA taking advantage of house lots, put it this way: "The more we grow, the less you mow." Hallelujah!

It was another standing room only event, but with a much different vibe than the previous occasion. While the first event had focused mainly on the agricultural logistics of farming by bike in the city—finding garden space, choosing crops, distributing produce—the second event was steered by the audience towards economic logistics—insurance, contracts, liability—all of which frankly bored me. In my opinion, the questions of how to make money or whether to embed oneself in The System could only be answered individually. That is, with urban farming—as with life—there was no "model" to make or to follow. So why get into the minutiae?

In retrospect, I can see what happened: the first year, the concept of "urban farming" was so novel that it was enough just to talk about what it was. The second year, thanks mainly to all the media attention, people wanted to mainstream it as a career. In this new phase, any "revolutionary" visions were at best a source of inspiration and at worst an expression of naïveté. The spirit of idealism was being eclipsed by the pursuit of cash. Not a new story, but I wasn't happy to see it.

I shared some of my reflections about the event in an email to the Sunroot Gardens list:

Date: Fri, 20 Feb 2009 12:51:36 -0800 (PST)
Subject: [sunrootgardens] 2/21 pea planting; Farm News

My biggest impression from Wednesday's meeting was seeing how many people are hanging on to old falling down structures and ideas despite the plenitude of evidence that they are on their way out. For people who still have faith in —for example—the insurance industry, I have one word: "Countrywide." That ain't a barn that'll be standing much longer, not with that lean.

The U.S. mainstream press is nothing less than the propaganda arm of the corporate elite, whether that is "officially" true or not. I don't care if that (or anything else I say) sounds like the ravings of a crazy person. Indeed, it is quite true that: "it is no measure of health to be well adjusted to a profoundly sick society."- J. Krishnamurti (1895-1986)

Do you feel like you "fit in" with the standards of Culture? That you are "making a living" in line with Society's requirements? That your relationships are healthy because they are based on an abstraction called "Commitment"? If so, then be grateful for such obvious red flags! It's easy to see what needs to be pitched! We all know these things about ourselves—the lies, omissions, and fakeries dressed up in pretty words and called "civility"—and no one is stopping us from letting them go.

I have found that such mind-trash is the biggest barrier between me and good farming. The plants don't give a shit about any of the people-culture crap; the only way to know them is to be present with them without anything between: without doctrine, without preconception, without intention. Only when the waters in my head start to clear do I begin to discover the living connections that are already there, among all the creatures. Nothing needs to made or "worked on." It is just my ego-mind that needs to get out of the way.

Clearly, my personal path was still leading me away from the center. Increasingly I sent messages like this out to the Sunroot list, and they met with varying degrees of receptivity. Sometimes people unsubscribed. It became a running joke for Angel to ask how many people had left the list after a particularly cutting message. We both considered the "real work" of Sunroot Gardens to be personal growth, not farming. "Spiritual" work, Angel called it (and still does). I didn't like that word (and still don't) but otherwise we were (are) in agreement; farming was merely, as the Buddhists put it, "carrying of water and chopping of wood": the daily labor you perform to feed and care for yourself. Hence our lack of interest in how to capitalize on it or make it fit into a society that we both found to be "profoundly sick."

All of the above work, personal and agricultural, appealed to our other new partner in 2009, Tom of Mall56. He lived in the house at the "Mall56" garden property and had watched the complete transformation of the lawn into a bean field in 2008, while he tended a small kitchen garden on its edge. In 2009, he wanted to get serious so we flipped the set-up, taking the kitchen garden for Sunroot and handing the bulk of the property to him so he could experiment at a larger scale.

In mid-April, Angel and I met with our two new partners, Demeter and Tom, to discuss the upcoming year. This account captures the excitement we felt for the present and the doubts we had about the future.

Date: Tue, 14 Apr 2009 20:32:50 -0700
Subject: [sunrootgardens] 50 Gardens (and counting)

Wonderfilled greetings from Farmer K:

On Saturday eve we held a Farmers' Convocation. Demeter of Demeter's Table, Tom of Mall56, Farmer K & Angel of Sunroot Gardens. Among the four of us we have 50 plots around Greater Southeast. Sunroot has 11 $-paying CSA subscribers, with room for 2 more shares (inquire with Farmer K if interested). Demeter has 8-9, possibly with room for more later. Tom is interested in market gardening, including the new farmers' market we have been invited to start on Hawthorne this summer.

From this network, harvest & distribution of produce will happen 3 times a week from at least 3 different locations starting June-ish. Farmer K plans to distribute to Sunroot CSA folks on WEDNES-DAYS from the Firepit. SUNDAYS will be the Hawthorne farmers' market. FRIDAY is for helpers, land-lenders, barter, etc., from a location TBA. Demeter will be distributing to her CSA folks on one or more of those three days.

By comparison, during the 2008 season, when Sunroot partnered with Melanie Plies' CSA, Backyard Booty, the network had 40 gardens, with 40 CSA households altogether, with distribution divided between 2 days. This year there are about two dozen CSA households, the number being lower so as to have more produce for everyone all around: subscribers, helpers, farmers.

THIS IS NOT A DRILL:

So this is quite an operation we've got going here. None of the four farmers own any of the land they are farming, and are sharing as much as possible. So we find ourselves witnesses to and perpetrators of new economies of gift and trade that are emerging below the radar of the crashing mainstream money culture.

One of these seasons—maybe this one—we human creatures in the industrialized West will find ourselves with a food system that is no longer providing enough for everyone. When that happens, the urban farming networks will be essential for staying healthy and alive. We farmers here in this network are playing this season like it is not a drill. Because we don't know if it is or not. So to treat it "for real" is the only responsible act.

History shows that changes in human social circumstances can be rapid. A typical city has enough food on hand total—on shelves, in warehouses, in trucks, etc.—to last less than a week. That's not much slack to play with if we experience interruptions in fossil fuels, drastic revaluations of the currency, or crazy weather, all of which are in the realm of everyday possibility. (Two years ago, everyone said you were crazy if you pointed out that the real estate bubble wouldn't last. Now, millions are in foreclosure.)

At the time, it seemed like the economy was teetering on the brink. It was, but the Federal Reserve took extraordinary measures to keep it afloat, i.e., the Quantitative Easing (QE) programs. Regardless, the teetering never went away and is still happening at the time of this writing, October 2015.

2009 was a year of innovation for Sunroot Gardens. Three examples follow, as announced in a couple of emails. The first was the offering of CSA produce shares throughout the Winter, unheard of at the time. Second, I reduced the number of CSA shares offered for sale, even though there was enough demand to increase them. The third was the addition of a pick-up day each week dedicated to barter customers and work-traders.

Date: Sun, 4 Jan 2009 19:59:39 -0800 (PST)
Subject: [sunrootgardens] Winter produce picking / CSA schedule

WINTER CSA SCHEDULE After some post-frost inspections, we have determined that we can deliver a weekly winter share for the months of January and February, bar any unforeseen circumstances (i.e., snowed-in again). So, every Monday in those two calendar months, produce will be set up at the Firepit Garden, for CSA folks to pick up from 3pm on. If you cannot make it on Monday, call me by that day and let me know to set aside something for you. In such cases, a cooler in a regular spot will hold such shares. You can expect to find me there until at least 6 most of the time, later than that if that (later than 6) is what happens.

Come March, with different growth habits, roots not holding the same anymore, etc., we will see what's going on, and whether the CSA will remain weekly at that time.

BTW, Sunroot Gardens is one of the few CSAs in the area offering a winter share. I believe we are the only one that is doing so every week. This is experimental. For being the lab rabbits, so to speak, you get sweet sweet carrots! Hope you're enjoying that.

Date: Sun, 1 Feb 2009 14:22:09 -0800 (PST)
Subject: [sunrootgardens] Early February Farm News

In 2008, Sunroot Gardens attempted to provide produce for 25 monetary subscriber households, plus helpers. Additionally, all the plots were being shared with another CSA, Melanie's Backyard Booty, bringing the total number of subscriber households to 40+.

This year, Sister Mel is working 1/3 acre on Johnson Creek for her CSA, and Sunroot has reduced the number of monetary subscribing households to 13. Sunroot will have at least as much, and likely more, real estate under cultivation in 2009. (New plots are still being added, so pass along any leads you have.)

Anything can happen of course, but given these numbers, more produce should be available for those contributing labor, land, goods/services in 2009 than was in 2008. Especially in the summer, when certain crops like snap beans, peas, summer squash, tomatoes, *must* be picked more often than once a week in order for the plants to keep producing.

Therefore, I am considering the idea of adding a second distribution day to the week, on Thursdays. This would likely not start until sometime in June. It would work like the Monday monetary subscriber day, where produce is picked, processed, and laid out farm-stand-style, for people to help themselves. Ideally, this picking day would be led and organized by farm helpers, who would take their own cut for participating. These duties could be rotated among a set of regulars.

In this way, farm helpers would have something regular to count on, in addition to produce/treats they receive on-site when helping on other days. The invitation to pick up regularly on this day throughout summer/fall could be earned by labor that needs to be performed before that, in winter/spring, such as greenhouse or bike-cart construction.

Officially opening up the CSA to barter in addition to monetary contribution in this way could make the farm enterprise more sustainable (to use that term meaningfully). This is also an enjoyable way of bending the CSA model in a time when it is in danger of becoming—as all things Americans touch—a staid, by-the-book, paint-by-numbers affair. In a world where economics are changing so fast—and the dollar could deflate or inflate into meaninglessness anytime in the next week to year—it makes sense to bend (or break) the CSA model if that's what will keep Sunroot going as a productive source of food for as many people as possible.

By some means of measuring—number of gardens, success of crops—2009 was the peak year for Sunroot Gardens. By others—participation by helpers, tobacco harvest—it was one year past. Vehicles intruded more than in the past but I got a custom-made farming bicycle, geared low for heavy loads, from a bike-head welder. On the farming front in general, I felt like a lack of focus led to more wasted time. On the personal ("spiritual") front, I found myself embarking into *terra incognita*, and in 2009, such harvests eclipsed the haul of vegetables.

02.08.2009: "Hairy Little Heart"

Date: Sun, 8 Feb 2009 18:57:52 -0800 (PST)
Subject: [sunrootgardens] Imbolc farm update

Imbolc greetings from Farmer K:

This week's featured green is an uncultivated Brassica Family plant often called "Bittercress." The name is not so accurate; it indeed has a "cressy" mustard flavor, but I find no bitterness to it I have been eating it raw when I find it for a couple weeks now, but cooked with for the first time last night. I heated it very briefly on top of cooked tempeh in a fry pan, with a splash of beer to steam it. This was served on quinoa. Quite excellent.

The botanical name of this plant is "Cardimine hirsuta." "hirsuta" = "hairy" and must refer to the slight fuzz on the stems, as the leaves are quite smooth. My guess is that "cardimine" means "little heart" and refers to the shape of the leaf at the end of each stem. So, instead of "Bittercress," we could say, "Hairy Little Heart." According to wikipedia: "Other common or country names include land-cress and lamb's cress. As Old English 'stune', the plant is cited as one of the nine herbs invoked in the pagan Anglo-Saxon Nine Herbs Charm, recorded in the 10th century." Just in time for Imbolc. ("Imbolc-cress"? OK, I'll stop now.)

Whatever name we use, the plant is an example of an edible plant that just comes up around the city. I am seeing it in places where we made rough garden beds from lawns last year, so it is a pioneer plant, thriving in disturbed areas. If left to go to flower and seed, its dry pods explode when touched, sending seed everywhere. Members of the Cardimine genus currently grow on all the continents except Antarctica, but I don't know if this particular species is "native" to this area or not. That is to say, if it came with

Europeans. No matter. I personally consider anything currently growing in this ecosystem without human cultivation to be "native," including things called, "invasive." They're native *now*, after all.

Trick Question: What percentage of the plant and animal species living on Hawai'i are invasive to those islands? Answer: ALL OF THEM. Hawai'i is a recently emerged land mass that stood as a barren mass of lava right at first. Everything now growing there "invaded" from somewhere. And so it is and always has been with all life on Earth, always moving around, thriving, going extinct, etc. The labels we humans put on things don't make a difference to any of those things.

The answer to the Trick Question is not totally factual. Some species are unique to Hawai'i because they have since evolved into fully differentiated creatures since their ancestors arrived.

03.03.2009: "Still Paying Taxes?"

Date: Tue, 3 Mar 2009 11:33:55 -0800 (PST)
Subject: [sunrootgardens] Sowing oats, getting busy, step up!

...Still paying taxes? You are acting against your own interest. The feds are giving it all to the super-rich, and the local governments are building prisons and arming their violent police forces to the teeth. The so-called educational system is simple brainwashing (yes even here with Portland's "good schools"), and "health care" is a money-making scam that injures, addicts, and kills hundreds of thousands of people each year. We are past the point where it's time to drop all that shit and to act radically different with our time/resources in order to survive. This is an unreformable system. A government that is oppressing its people is *not* aberrant; it is doing exactly what governments are intended to do: Conform and control. But everyone can escape it. Great personal freedom lies outside of that mess. Go ahead and make the jump Outside.

There's already people here, waiting to welcome you. You can do it no matter what your circumstance. Debt and/or children are no excuse for staying Inside. In fact, they are two of the best reasons to bail yourself out.

The rubber has hit the road, folks. Our efforts are not theoretical or ideological—they are survival based. If we are to eat, we need to work on it now. C'mon, let's go!

04.26.2009: "Farmers without phones" / "What's going on"

I am including this longish email in its entirety because it captures so well the tenor and circumstances of the time: Despite relative financial impoverishment, I was not suffering; the collective focus of the farmers was on the work at hand; and the opportunities being offered to helpers were unique in their structure and scope.

Date: Sun, 26 Apr 2009 10:55:25 -0700
Subject: [sunrootgardens] Farmers w/o phones / Spring Planting / Produce & Staple Crops shares

1. FARMERS WITHOUT PHONES
2. SUMMER VEGGIE / STAPLE CROPS PLANTING MADNESS
3. AVAILABLE SHARES IN CSA / STAPLE CROPS
4. "I Want To Help—How Do I Do That?"
1. FARMERS WITHOUT PHONES

The waning of the last moon (which just turned new on Wednesday) brought Farmer K a broken cell phone (transplanting accident), a busted laptop (falling box incident), and a munged bicycle (normal use). 'twas an interesting place to find myself, with the three tools unavailable at once. The bike was the most missed item, but a cycle-head farmhand made it ride-able again by shortening the chain and bypassing the crunched dérailleur, giving me a functional one-speed ride. Fine for getting around, not so good for hauling loads. And I finally have one of those cool bike chain bracelets, made from the extra piece cut out.

A replacement phone arrives today. Corey at Seven Corners is finishing up TWO farm bicycles, to be ready this week, and the laptop will likely recover with an Ubuntu 8.x start-up disk & reformat.

Demeter, of Demeter's Table, one of the Four Farmers of the PDX Apocalypse, is also cellphone-less at the moment. She has been doing the buy-minutes-deal, and is seeking a longer term solution. This is an opportunity for someone with a "family plan" for their cell service to add a phone for her use in exchange for a CSA share. This would be a great $ deal for whomever, as the cost of another phone on a family plan for X months is less than the total cost of a CSA share. SO—write me or Demeter if you would like to make this offer to her. It certainly is convenient for us farmers to reach each other by phone when we are working on 50 plots around town!!

2. PLANTING FOR SUMMER VEGGIES / STAPLE CROPS

The moon is waxing. It is full on May 9th. The next full is on June 7th. During these next two cycles, beginning NOW (as in TODAY), here is what must be planted, if we'd like it later:

* All squash—cucumbers, summer squash, melons, winter squash for storage
* All beans—snap beans for summer, soup beans for winter. PLUS, soy for summer edamame and winter miso
* All corn—sweet for summer, flour/dent for winter
* QUINOA—for survival
* Sunflowers—for oil, for cooking

So here it is. For STAPLE CROPS, what we plant in the next 6-8 weeks will be what we have to eat over this fall, winter, spring, 'til next harvest next year. This is a vital time, if our concern is to truly take care of ourselves. For 10,000 years, most of the human creatures in the world have been eating by growing & cultivating, harvesting & processing, and storing. Just exactly what we are talking about with the Staple Crops Project. For the last 50+ years, most of the human creatures in the Industrialized West have been eating by buying it from shops. The food in the shops has been grown and transported and stored by using a one-time-only supply of concentrated solar energy stored in rocks and viscous liquids (coal, oil, etc.).

It is now clear, to anyone who cares to notice, that this 50+ year system is running out of steam, and—as in the rapid phase changes found everywhere in nature—could give out rather suddenly any time now. We don't know when. Many people say, "Oh that's years away." When I ask these people for evidence of this claimed longevity, they have nothing to offer. No, what these people are talking about is not a reasoned and measured examination of the facts and of their possible outcomes; what these people are talking about are their own personal hopes and fears about the system. Well, I have to tell you now, I don't give a shit about anybody's hopes/fears, including my own. When it comes to EATING, which most people currently have to do, we must look at the FACTS of our situation. Our FEELINGS are nobody's business but our own, and have no place in a practical discussion. And the FACTS show that some kind of "Crash" is not theoretical, but is actually already occurring and has not reached us here yet. One billion human creatures already don't have enough to eat. That number is rising, and will include more of us here in the West, sooner or later, regardless of whether we fear it or hope against it.

So, all that being said, we are planting like crazy over the next two moons here. If necessary, we can LIVE off of quinoa, beans, squash, winter vegetables, etc. We have the seeds, we have tools, we have amendments, we have land. All we need is more folks to come out and help. So how does that work? How do you get in on this food? Read on....

3. AVAILABILITY OF PRODUCE / STAPLE CROPS

For those who have little/no time, and want produce or staple crops, there are a limited number of shares available.

* CSA $ SHARES: Produce shares with Sunroot Gardens are $600 (negotiable). This is for 30+ weeks of produce (more than most CSAs), starting in May, and going through Winter Solstice. Pick-up will be at the Firepit on Wednesdays. There you will find the week's veggies/fruit spread out like at a farmers' market stand, and can take what you need for your household. The total number of CSA shares is being kept down to 13 in order to assure a good amount of produce for each household. This is down from 24+ last year. TWO CSA SHARES ARE AVAILABLE.

* STAPLE CROPS $ SHARES: These are $420.11 each. (That's the year's budget divided by 10.) Each share gets you a fixed percentage of the final harvest. This will translate into hundreds of pounds altogether. Certainly enough for a household (or two or three) to make it through the winter if necessary. Crops will include

wheat, oats, flax seed, quinoa, soup beans, cooking oil (hopefully) and flour corn, plus more. TWO STAPLE CROPS SHARES ARE AVAILABLE.

* The NO-$ approach!! Sunroot Gardens will have a second veggie/fruit distribution day every week, on FRIDAYS, for people who donate labor, land, goods/services. This will be by-invitation. Again, it will be set up like a farmers' market, with piles of produce for you to pick from for your household. Additionally, people who come to help will be offered whatever produce is available on the days they show up. In some cases, helpers get "special" produce that is not growing in enough quantity for the Wed. & Fri. picks. The famous salmon-colored raspberries of the Firepit Garden are one example. "The best raspberry ever," Farmer Angel says.

STAPLE FOODS LABOR - There are TWO ways to get food from the Staple Crops Project besides $. One is to be a HELPER. Hours worked by helpers are logged, and the helpers will receive a percentage of the harvest proportional to the amount they worked. Last year, folks who helped harvest and/or process the wheat got one pound of wheat berries (or flour) for every hour worked on the project. This year, that amount per hour is expected to be higher due to more efficient harvesting techniques. If a household sends three people for three hours, that counts as nine hours total for the household. To work on the Staple Crops Project this way, you just show up for those days when you want.

The other way to get stuff out of the Staple Crops Project is by becoming a FARMER on one of the projects. The total harvest is divided among a) $ investors, b) time investors, and c) "the farmers." FARMERS are folks who are "bottom-lining" a particular crops somewhere. For example, someone with an interest in soup beans might volunteer to "bottom-line" the Cannellini bean patch, which could be up to 1/4 acre in size. "Bottom-lining" does not mean you are working alone. What it means is that your own personal interest in that crop is strong enough that you will help find materials and tools, that you will go around and check on it regularly, that you will help organize work parties for thinning/ winning/cultivating/harvesting/processing, etc. What FARMERS get out of this extra work is a share of the Farmers' portion. This is expected to be a greater amount than you would get just by working hours as a helper, since the number of farmers on any given project will be limited. The farmers' share of a crop is intended to be MORE than the farmer's self needs, so that the farmer can barter/sell some of it for other things in life.

Whew. So that's a description of what's going on currently with Sunroot Gardens CSA & Staple Foods. If you would like to be involved with any of this, contact Farmer K at 503.686.5557, or write back to this email.

4. "I Want To Help—How Do I Do That?"

STEP RIGHT THE FUCK UP. That's how. This email is about all of the "courting" that I can give to anyone. Yes, there will be "work parties" for big projects, but in the meantime there is LABOR to do every single day. Yes, EVERY SINGLE DAY, ALL DAY, ANYTIME you want to come help, we have something to do somewhere. Call/Write one of the farmers to find out where things are happening that day. The numbers are below.

I, Farmer K, am now basing day-to-day operations out of the greenhouse at the Firepit Garden. So you can also stop by there anytime to see what's going on. This time of year, we will be "suiting up," loading tools, etc., there in the morning for the day's activities. If you show up early enough you'll get breakfast cooked by Mrs. K. She also serves dinner there to whomever is around at sunset. It's a lot like camping, but with gourmet eating.

On one final note: In my mind, the entire point of helping around the farm is to ENJOY YOURSELF. The world of the gardens and our communal resourcefulness is a haven from the exhausting work/entertainment complex of city-life. In the gardens, you are invited to set aside your socialized fears & doubts, and to find presence with the life of the plants and soil and water. No experience is necessary. There is very little to KNOW, but there is much to NOTICE. What is being offered here is an opportunity to find moments of true integrated living. We all know that Society is based on non-material abstractions ($, "ownership," etc.) and many people feel trapped by those things. The garden is where you can step out of that silliness and experience something REAL instead.

C'mon out!!

—"Human culture is the attempt to deny ourselves and others the most basic truth of our existence: that we are absolutely free and incomparably perfect, every moment." -KtS

05.25.2009: "Labors & love happening in multiple locations every day"

Another email that illustrates the frenzy of planting season.

> **Date: Mon, 25 May 2009 22:48:11 -0700 (PDT)**
> **Subject: [portland-urban-farmers-coop] FARM NEWS**
>
> FARM NEWS from Sunroot Gardens: Being a compendium of what's seeded, what's bolted & gone, what's sprouted & on the way, what's weed-whacked, and what's still just a dream...
>
> Due to weather conditions, things in the agricultural world are about 10-16 days behind this spring. (Put another way, the Gregorian Calendar, promulgated by a Pope several centuries ago, is not lining up with actual conditions this season.) Orchardists in the Gorge are saying the fruit is a week and a half behind, and that cherries will be too late for July 4th, their traditional big opening week. Other farmers I have spoken to are reporting similar stories. I myself have been looking around at the gardens the last week or so and saying, "So where's the food?" I'm still subsisting mostly on over-wintered leftovers that are becoming tougher and harder to come by (though with the cooler spring these overwintered greens have produced more and later). Most—as in 4/5 or so—of the brassicas we have seeded this spring have bolted prematurely. Everything from broccoli raab to mizuna to radishes to turnips to mustards—BAM!—shot up flower stalks and turned stringy, in some cases w/in a 48 hour period.
>
> Many summer veggies are coming along strong because of the extra work done with greenhouses, cloches, cold frames, etc., this spring. For example, there are 30+ tomato plants under a tunnel cloche at the Firepit, sized up good for this time of year, with their

own irrigation, and trellising ready-to-happen. At Ninja Garden, 60+ summer squash plants are plunked into mounds and putting on their 3rd and 4th leaves. They were starts in the greenhouse at the Firepit first. Cukes and basil are already started and being moved outside as appropriate. A couple hundred cabbage starts wait at the Firepit for a bed to be cleared for them at Soleil. Winter Squash starts are being hardened off at Chiens, to go to plots in outer SE and Sellwood, along the outskirts. And at Otis's Waggin-tail garden, over 500 row feet of bush beans have been seeded. Nine different varieties including French fillets in green and yellow, Romanos in gold, green, and purple, and Snaps in gold, green, purple, and red. Then there will be pole beans, including Asian yard-longs, in several colors and shapes.

On the Staple Crops front, we spent three days at the Hampton acreage in Milwaukie and seeded quinoa, flax, soldier beans, French horticultural beans, dent corn, Cuban tobacco, buckwheat, millet, parsnips and amaranth. Plus a mad mix of seeds intended to be an insectary blend, to attract beneficial bugs. Of the crops mentioned above, quinoa is the main one on the property. We put in 15 or so 300-foot rows of quinoa in one section, and broadcast it in another. Recent visits to the Carver acreage has shown a 1 1/3 acre wheat crop doing very well, especially where we gave it AZOMITE. The other 2/3 acre available to us there has oats, fava beans, flax, oilseed sunflowers, cover-crop radishes, and a little quinoa, all coming up and doing their thing.

The multi-farmer collaboration continues, with gardens being sliced, diced, traded, flipped, and otherwise co-stewarded by whomever is there to do it. People who come along are invited to take on as much or as little as they would like, and that's been happening. For example, Tom of Mall56 got to drive the tractor to-day at Hampton, and seemed to get quite a thrill out it. The tractor is Angel's toy. It's a 35hp Massey Ferguson. Quite a toy that thing is, and you need to ask *him* not me or anyone else if you can play with it. There will be other chances, along the Staple Crops way.

To join in the fun frenzy that is spring/summer planting, simply con-tact any of the farmers listed at the end of this email. Labors and love are happening in multiple locations every day now.

07.10.2009: "The Hobbit Meal Plan"

Date: Fri, 10 Jul 2009 18:19:15 -0700
Subject: [portland-urban-farmers-coop] Sunroot Gardens
FARM NEWS, incl. "The Hobbit Meal Plan"

...here at Sunroot Gardens, we have been observing the Hobbit Meal Plan, which is SEVEN meals per day:

* Breakfast
* Second Breakfast
* Elevenses (or "Elevensies")
* Luncheon
* Afternoon Tea
* Dinner
* and, late in the evening, Supper

With 16+ of workable light per day, seven meals is honestly a minimum. None of these meals is generally very big, and can be as simple as an Americano and a pastry in exchange for some produce at a local coffee shop.

Sunroot Gardens has been providing as many of these as possible for whoever is around helping that day. The farm invested in hundreds of pounds of bulk food (rice, quinoa, polenta, oats, various beans, cooking oil) last fall, with the idea of having enough to feed hungry farmies until the Staple Crops harvests come in. So we're still running through all of that.

We also often get fresh corn tortillas, still warm off the tortilla press, at Yesenia's Supermercado on 67th & Powell. They are delicious, versatile, and a mere 75 cents per pound. They also sell fresh-style cheese ("queso fresco") there, made in Hillsboro, at only $5/lb. That's right, local food that isn't wicked expensive, like at the snooty gringo stores (i.e, the co-ops, New Seasons, etc.).

Digging into some home-cooked beans with tortillas and queso at Luncheon, after hours in the field, with hours still to go, it becomes obvious that farmers had a part in the creation of these kinds of foods: easy to eat without elaborate utensils, fueling-up without over-filling, tasty and easy to dress up with whatever produce is local. Asian farmer food can also come into the diet in the form of sticky rice balls, with beans/veggie fillings. We enjoy simple deliciousness and fortifying nutritiousness in these workday meals. Then there's surprises, like at the Claibourne Estates Garden yesterday, when owner Rob invited X and I into the house for French Toast for elevensies with homemade strawberry jam and pomegranate molasses. "LIVING LARGE" we declared it. And indeed, we are eating better than most people in what is called "the united states," including the moneyed-class. The trumpeted advantages of conventional employed insured licensed work/play ring hollow in the face of actual simple fresh present undomesticated living. And the culture at large is at the beginning of a transition from the former to the latter, so we are happy to be taste-testing the future thusly for all y'all !! ;-)

This particular email was epic in length and contained another section I wanted to include here because it shows how different Sunroot was from other CSAs. The standard CSA arrangement is that the customer is presented with a box of food with carefully measured-out amounts of each item available that week. At a Sunroot pick-up, however, all the produce was set out "market-style" and subscribers were invited to take as much or as little of whatever they wanted. I made this clear to everyone as they signed on and paid up but it was still confusing for people sometimes, so here I laid the concept out again (for what felt like the umpteenth time).

....Reminder of "how much to take" on CSA day.

The short answer:

WHAT YOU WANT

The long answer:

The complaints about CSAs that I hear the most are along the lines of "I got too much of X, not enough of Y, and I just don't like Z." That is, most CSAs divvy up the harvest into equal parts and put it in boxes for people to take. At Sunroot, this is not the case (except when explicitly set forth for a particular item, as in "one each of X this week"). The produce is set out in piles, like at a Farmers' Market, with the idea that YOU take as much or little of everything. ONLY ***YOU*** KNOW HOW MUCH YOU WANT!! Do NOT worry about "leaving enough for other people." That is MY puzzle, as the farmer, NOT YOURS as the CSA subscriber. After all, how will I ever know how much to pick if people aren't taking what they want? It might be "polite" to not take what you want, but it is certainly not Thoughtful or Practical. "Politeness" has no place in farming. Thoughtfulness and Practicality do. I am fully aware, partially through these CSA experiences, that many people in this society are uncomfortable with even getting into touch with what they want, regardless of taking it, but that is not my problem. CSA Pick up with Sunroot is an opportunity to put all the abstract civil nonsense to the side and simply address your desires on the spot.

The middling answer:

Other CSAs just give you something and say be happy with it. With Sunroot, we want to give you what you want. The only way for that to happen is for you to show us what that is.

In-Depth — The Versailles Syndrome: All About Landowners

Every garden we tended was on land that someone else owned. So to one degree or another we did not have full control over any of them. That's difficult for a farmer. While it's true that rural farmers who own their own land also face restrictions—whether legal (land-use laws) or social (keeping peace with the neighbors)—no one can tell them what to plant, where to plant it or how to tend it, for the most part. As urban farmers, we had people breathing down our necks about all of those choices and more. What could we do? We were on *their* property, after all—"at their grace," as it's said—so the best we could do was to try to choose wisely. Sometimes we did, sometimes we didn't.

Because some of the people we worked with were renters, not owners, we called them "landlenders." Potential new landlenders always asked, "How does this work?" and we always answered, "However you'd it like to." We told them that each landlender arrangement was different. Some received produce, others didn't. Some allowed us to use the water, others forbade it. Some demanded a written contract, others were satisfied with a handshake. No one, interestingly, ever asked for money. Nearly all of the landlenders approached us first; we rarely added new sites by walking up to strangers with a pitch.

During the time I farmed in Portland, I had relationships with at least forty different landlenders. As could be expected, these relationships ran the gamut from truly enjoyable to downright infuriating. It was kind of like dating: Fevered excitement in the beginning often led to disappointment, and many relationships were terminated by one or both parties after only one season. Other matches were decent and grew in fruitfulness for everyone over time. In a handful of cases, intimate bonds were forged. (Here's looking at you, Lady Quince!)

Because we were farming in so many locations at once, no particular spot was ever essential to our overall success or failure. This gave us a wide latitude of freedom: No single landlender could ever hold everything over our heads. We could (and did) drop gardens as we thought best. Because of ongoing media attention, offers of new plots were always coming in.

But, because we were farming in so many locations at once, few gardens ever received the full attention they deserved. With each season, we became more creative at setting up plantings to require as little maintenance as possible, but it was an uphill battle. The agricultural issues raised by a lack of daily attention were challenging enough, but the social issues, also unavoidable, were far less of a joy to tackle.

The most common point of contention with landlenders was aesthetic. Farming doesn't *look* like gardening, and our farming especially didn't. Conventionally, the ideal city garden is well-manicured, weed-free, and set out in neat little rows. Of paramount concern is appearance. Conversely, the number one priority of farming is productivity; if a bed is a little rough-around-the-edges but it's pumping out big harvests, then the job's getting done. We were not *un*concerned about the appearance of our gardens, but we were (usually) unwilling and (often) unable to perform labor we considered unnecessary. Of course, when a landlender got persnickety enough, then the unnecessary became necessary just to keep the peace (and the plot).

But beyond questions about whether the cabbage patch is "neat and tidy" enough or not, farming also includes certain practices that—by their nature—can look unkempt or even chaotic to the untrained city eye. Two examples that repeatedly raised issues for landlenders or their neighbors were cover-cropping and seed-saving, both central to small-scale, sustainable agriculture.

Cover-cropping is a method for creating and maintaining the healthy soil needed to grow nutritious vegetables. Depending on what is planted when, cover-crops serve different functions; it can add nutrients, smother weeds, create green manure for soil-building, discourage pests and protect the ground from compaction and excessive mineral-leeching during the rainy season. Cover-crop seed is usually inexpensive and the plants themselves generally require little care, so it is an affordable and fairly easy way to improve land for vegetable-growing.

But a stand of cover-crop does not fit the picture of the ideal city garden. The seed is broadcast thickly to raise a dense stand, so there's no rows. The plants themselves are alien to the urban landscape and can send a "wrong" message: oats, rye and wheat taller than a few inches look like un-mown lawn grass; vetch is tangly, buckwheat is gangly, and crimson clover is a jumbled mess. To make matters worse, a blend of cover-crop plants is often grown together and can appear disorderly (which, honestly, it is). Basically, a vibrant patch of cover-cropping resembles a thick patch of tall weeds if you don't know what's going on. Some people, including a few landlenders, were aghast at the appearance of cover-cropping and a more than few of our patches were weed-whacked down prematurely.

Here, in an email to the landlenders, I offer reassurance about the appearance of their properties and some education about cover-cropping:

Date: Sun, 06 Jun 2010 13:29:22 -0700
Subject: [sunrootgardens] Save Mall56 Garden! / What's Growing update / Future Q's

Note to Landlenders

Nope, we haven't forgotten about farming in your yard. Very rainy conditions have been preventing any labor that involves working the soil. Soil structure can be damaged by messing with when it is too wet. Best just to leave it alone. We are aware of the passing of days, of the lateness into the season. We have seen the non-farming folks edging their lawns, chopping down their "weeds," digging up beds, and plunking out supermarket starts because it is the Calendar Date for such things. As farmers, we know the calendar dates never exactly match up to a given year's agricultural cycles, and that we must respect the conditions caused by weather if we are serious about getting good food out of the ground. "Stewardship" is a word one could use here, in which we strive to avoid knowingly abusive treatment of the ecology we are attempting to partner with.

BTW, we have noticed—by comparing dozens of locations over the last few years—that yards that have been under a cover-crop for some portion of time are much easier to work, and earlier. City-folk aren't used to seeing 3-foot tall fava beans by their driveway, or 2-foot crimson clover, or 7-foot rye, but these practices lead to better soil tilth and health. At one particular garden, we can show you side-by-side beds in which one was cover-cropped over the winter, and the other had overwintered kales/collards. Cabbages are now planted across both beds, and those in the brassica section are being ravaged by slugs, while whose in the cover-cropped section are mostly fine. It is a dramatic difference. The overwintering brassicas made a great winter home for slugs and slug-eggs, while the wheat/pea cover-crop did not. Cover-cropped beds also dry out quicker in the spring because all that plant matter is pulling up moisture and transpiring it into the air.

By late July, when everything is lush and beautiful and productive, you will have forgotten the weediness or bare muddiness currently on your property. We appreciate your understanding !!

Though cover-cropping raised some ire, the most offensive act of farming we committed in the city was seed-saving. If cover-cropping was a rude remark, seed-saving was a slap in the face. Here was one email to the landlenders about seed-saving:

Date: Mon, 26 Apr 2010 09:18:44 -0700
Subject: [sunrootgardens] Beltane Produce & Bonfire / Updates for Staple Crops, CSA & Landlenders

Beltane Greetings from Farmer K!

Sexy Landlender Note

This year we will be redoubling (or quadrupling) our seed-saving efforts, including some work for a regional seed company, who gave us seeds in exchange. There's a phrase in English, "gone to seed," that has negative connotations. The reckless or even rakish appearance of a brassica gone bolted and drying out in the August sun sure can look like a weed to the untrained, urban eye. Indeed, we have been accused of "not taking care" of plots because they were in full flowering glory. As is commonly known, sexual reproduction can be a messy, disheveling affair, and it might be yet another shoot from the culture's Puritan roots that even plant sex is somehow improper to witness in public. For chrissakes, as Holden Caulfield says.

So anyway, if you gaze out your window to a plot of fuzzy-headed lettuce towers, or gangly twisted chards, or towering carrot umbels, just think of them as having "bed head" and let yourself be

lustily inspired by the spray of petals, the sultry hum-buzz of polli-
nators, and the tumescence of expanding seed pods....

This was a humorous attempt to address an issue that was very serious to us. Every farmer wants vegetables that are hardy, delicious, and easy-to-grow, and for that the farmer needs seed that is strong, well-selected, and locally-adapted. Much of the seed on the market, however, is of low quality or unknown origin. Even reputable seed houses sometimes ship bunk. Additionally, we are now in the era of Genetically Modified Organisms (GMOs), facing unknown but potentially catastrophic dangers.

Seed-saving is the ultimate way of knowing where your seed came from, of course, and it is also an essential step in improving a vegetable for hardiness, etc. Basic plant-breeding is easy to try. Without doing anything special, the first generation of saved seed from a store-bought packet is already a meaningful "selection" because it grew from that percentage of the purchased stock that was a.) viable and b.) able to reach to seed-bearing age locally. In following generations, further selections can be made for appearance, flavor, shelf-life, etc. Seed-saving is an essential activity for the small-scale farmer, including the urban farmer.

But, as noted above, there are negative social implications to letting a plant "go to seed." When compared to the ideal city garden aesthetic, a bed full of bolting lettuce, spinach, or mustard greens appears untended or even abandoned. Over our years of urban farming, we lost seed crops in multiple locations when people cut down patches before they were ready. Nothing got the weed-whackers revved up like a patch of vegetables in flower.

Arriving at a garden to find beautiful plants chopped down was frustrating, infuriating and sad. Frustrating because a crop had been wasted after months of tending, infuriating because the cause was human stupidity, and sad because companions in our lives had been killed. I can't count how many times it happened, and it was often perpetrated by the landlenders themselves.

All in all, I would guess that many of landlenders were dissatisfied with their relationship with Sunroot Gardens, at least sometimes, and a few of them most of the time. In the first couple years, we were partly at fault for this by setting expectations too high. But later, when we tried to be more realistic, we still displeased people with our bountiful creations; they just didn't fit the picture that people had. Most of them wanted some miniature facsimile of Versailles—or at least a well-ordered *potager* garden—but what we delivered was the provincial peasant farm that the nobility screened off with high hedges.

If the landlenders had done as I had, and visited the rural operations of their favorite market vendors, what they would have seen would undoubtedly have shocked them: farms, as a rule, are messy, and the exceptions generally betray misplaced priorities. The vegetable beds are never entirely weed-free; piles of pots, plant trays, and plastic sprawl where they were last used; broken equipment waiting a day of repair that might never arrive are half-obscured in vegetation. The barn door is hanging from one hinge, the hoop-house covering is torn, and the irrigation leaks. The list of "I'll get to that next" projects is usually long. By the standards I had seen we were actually quite tidy, but the landlenders didn't know that.

Some of them, of course, were appreciative and totally pleased with us. Because those people were the most pleasant to work with, their plots got the most attention. The old saw, "the squeaky wheel gets the grease" didn't apply; in our world, it got the boot.

By 2010, when we were winding down Sunroot Gardens, it was abundantly clear to us that most landlenders were not ready for farming in the city. And that when they finally did feel ready, it would probably be too late.

09.02.2009: "Invitation" (to *Terra Incognita*)

At times, the emails that I sent to the Sunroot list were not merely personal but out-and-out confessional in nature. This one dives into that territory after a bit of farm news. I knew there was a set of subscribers who loved it when I went there. Others not so much, so this particular missive included the instructions for unsubscribing right at the beginning.

Date: Wed, 02 Sep 2009 14:20:57 -0700
Subject: [sunrootgardens] Farm News / Invitation

Greetings to all, Farmer K here:

If you want to unsubscribe to this list at any time, simply send an email to: sunrootgardens-unsubscribe@lists.riseup.net

In FARM NEWS:

1) Summer squash have finally run down. This is the last week they'll be offered.

2) Tomatoes still green. I believe I over-pruned them at one point, so they are behind everyone else's. We'll see.

3) Beets beets beets: I had no idea so many people like beets. They've been popular all around, and are sizing up on regular basis right now.

4) Greens: Soon we will be offering lots of baby greens. They will be thinnings from the massive beds of kale, mustard, turnips, and other brassicas that are coming up for Winter.

5) Winter Squash: Got more of those coming than in any previous season. 6-8 varieties. The leaves have been dying back from the powdery mildew that typically hits squash family here in Cascadia.

These include pumpkins.

6) Medicinals: We are entering the season to begin harvesting many of the medicinal herbs that are growing around town. The plan is to harvest and tincture as much as possible. If you are interested in tinctures, let me know—Assistance with purchasing grain alcohol is needed. And if you want to learn how to harvest and process medicinal herbs, in exchange for finished product, just call yourself out.

7) Basil / Pesto: We've got lots and lots of Basil growing around. If you are interested in making pesto, especially if you want to share some with Farmer, let me know and we will pick you what you need.

8) Nicotiana: The Firepit is hosting what I would wager is the broadest collection of Nicotiana species in the bioregion. At least 15 different species grace beds here. "Nicotiana tabacum" is what is contemporaneously considered "smoking tobacco," but the others also contain nicotine, to a less or greater degree. I broke a 20+ year corporate tobacco habit by switching to homegrown tobacco, and I can feel the difference, physically and otherwise. If you are interested in exploring this option, let me know. All tobacco products are given freely, with no expectation of return.

In OTHER NEWS:

The heat wave of a couple moons ago was an intense experience for me. I had not experienced such temperatures in many years. For a few days there, all I could do was sit very still in front of a fan in the shade. I hosed myself down with water every 45-60 minutes until dripped dry, then repeated. With no physical activity possible, and with the solitude that marked those days, I found myself turning within. What I can only call a "phase change" in myself seemed to occur under the stress. One day I started crying, which is something I had been unable to do alone in many many years (perhaps the Spring of 1987 was the last time).

The heat passed on, but things had fallen away that did not return. The opening did not close. What I can only clumsily describe as some sort of "expansion" has continued in the time since. I don't know what that means, but there is definitely nothing for anyone to feel about and no reason to worry.

As this moon has been waxing toward Full, I have been experiencing Sorrow to depths indescribable. I have been crying harder and longer than I can remember ever having done. Here is some of what has come to me during these sessions:

1) Self-consciousness: specifically, fear of what other people within earshot might think, has sometimes been a barrier to letting the crying start in the first place. This has been fading.

2) No Back-To-Normal: When one particular crying jag was winding down, I reflected on the fact that people, myself included, often call it "getting back-to-normal," in reference to when the tears have stopped, and one is breathing calmly, etc. This is the view that crying is an aberration; that is, a "going astray" or "wandering away." But in that moment, it occurred to me that "There is nothing to get back to." That is, the tears did not take me away from anything; if anything they brought me closer to something. If there is a Self, it is nothing to "get back to"; it is merely something to experience as it is unfolding, which includes everything that "happens": tears, laughter, silence, etc., alike.

3) Nothing's "wrong" / The Perfection of Existence: Many of the times that I have cried, I have haunted myself with thoughts like: "what's wrong with me?," "why can't I let this go?," or "what a stupid reason to be crying!," etc. These sorts of thoughts are examples of what is classically called "Suffering." It's really just a basic non-acceptance of the moment's reality; a disappointment that one's current experience does not match up with one's expectations or hopes. During one particular cry, after a sleepless all-nighter, I was there in that moment of "suffering," when suddenly it disappeared. I felt, deeply and without reservation, a full acceptance of my state of Sorrow. I felt, in fact, to be in a state of "Perfection," in the way that a circle is "perfect" in its entirety and wholeness, without flaw. Two brothers happened to be there with me, offering me their company, and I said aloud, "I am perfect, and so are both of you." Then the tears that poured from eyes were tears of Joy. For hours afterwards, I moved through the day with a crispness of vision, filled with the wonder of just being alive in a body.

4) There is no "Narrative" / Nothing to cry "about": What "made" me cry on any of these occasion? In most instances, there was some experience of inner suffering that preceded the tears. However, once I was crying, I would find that the particular incident in

question was not what I was crying about anymore. Instead, I found myself merely in a State Of Sorrow, the same way one can find oneself in a state of Love or Bliss or Anger. I am not here going to attempt to define any of these states, but I will say that I find the way they are defined by conventional society to be useless. Society sees them as things to seek out or to avoid, but that's not how they work. They are energies that pass through us (or that we are passing through, if there's a difference) and we perceive these energies in different ways depending on our current state of consciousness. Perhaps they are all just one energy, truly undifferentiated, and our interpretation of them into these classes, desirable and un-, is merely the muddle of confused minds. We are, after all, living in a poisoned world, with our bodies and brains under constant assault. Any way it is, I have found that there is no "narrative" here; no "story." "What are you crying about?" is, in other words, not a meaningful question. "It'll be alright" is another piece of nonsense. Sorrow is just one state in an everything that is always alright. (Not that I will necessarily be present in that way the next time I cry. We'll see.)

5) Beauty: One brother who was companion to me during a particularly cathartic crying session remarked on how beautiful I was just then. But this beauty that he saw was merely a reflection of his own beauty. The door that opened to me to see my own Perfection in a moment of Sorrow went wide enough to let him through, too, apparently. This is real Intimacy (a word that comes from the Latin for "innermost"). The "intimacy" that Society refers to, as in "men are afraid of intimacy," is not the same thing. *That* "intimacy" is generally an exercise in ego-construction, involving the fetishization of feelings, and then the displacing of responsibility for those feelings from oneself onto others. That's an emotional Police State, is what that is.

6) Don't Worry: So, long story short, there is no worrying to be done about any of this. I am not endangering myself or others, as they say, and I don't need things to be different. In fact, this morning, after a 20-minute tear fest, I wandered out in to the garden to hook up a hose on the carrots, and found myself completely filled with a sense of wonder and gratitude that I have been having these experiences at all. The "break," as it were, with a generally less-crying self, seems a great gift. I asked: How could the world be so beautiful as to put me in this place? And yet it has. Joy, what joy.

7) You're invited: I am open to sharing these spaces with anyone who wants to open themselves to them too, or who finds themselves already so opening. This is not an offer to instruct. I have nothing to teach. But I am intrigued by the sensations I have enjoyed of "going" to these places in companionship with others, as has happened of late. I am not talking "heart circle" shit. I am talking about "terra incognita"—the "unknown territory"—where there are no answers because there are no questions.

I close with the following Jiddu Krishnamurti quotation: "It is no measure of health to be well adjusted to a profoundly sick society."

Exactly one person replied with interest in my "Invitation." Though I never followed up with her, I appreciated that she had said something.

09.17.2009: "The Farmer Life"

Tears were not all I shared on the Sunroot email list. Joy, too, found a place, as seen in this message, though its expression follows a biting cultural critique. The harvests of my personal ("spiritual") journey were as important to me as the harvests from the gardens and I offered them to my subscribers just as I did vegetables.

> **Date: Thu, 17 Sep 2009 02:21:59 -0700**
> **Subject: [sunrootgardens] Wheat update / Plums galore / The Farmer Life / $**
>
> ...As I sit here typing, I am grateful to see that the sky is clearing and that the rain is passing along, apparently for the night. I am also grateful that there was no wind and that the rain was falling straight down. This means that my bedclothes under the market tent in the back of the Firepit are likely not (too) wet. When the rain started I went out there and made sure that the wool blankets were completely covering the cotton sleeping bag and the cotton shirt I put on the rolled coffee bag that is my pillow. I have noted that, even when the rain doesn't come under the tent, that anything cotton gets damp anyway, unless covered by something wool. Wool is warming even when damp (or soaked). I have confidence, then, that I have a dry place to cuddle into whenever I finish this thing.
>
> The rains are coming closer together these days, and the nighttime temperatures are starting to fall. At some point it will be too wet to sleep out there, and I will crash on the floor in the greenhouse. With that arrangement, I will need to roll up my bedclothes in the morning so there's somewhere to walk. Within 6-8 weeks, nighttime temps will be so low, and the dripping in the leaky, uninsulated (and uninsulate-able) greenhouse will mean that I will

need to find someplace warmer and drier. Last year's solution—the 14x8 trailer known as "The Winter Palace"—proved to be too expensive to heat, so we will see what comes up. It was not until October last year that I knew where I would be staying, so I am not concerned right now....

Living in an "inside" space with other people, like in a house or apartment, even if rent was not charged, seems an unlikely scenario. Having been "outside" for a year-and-a-half, I now see "inside" for what it almost always is: A set of dull, lifeless boxes containing dull, lifeless people creating dull, lifeless relationships based on dull, lifeless conceptions formed by dull, lifeless perceptions. You know how animals behave when put in cages—they become physically unhealthy, emotionally unstable, and aggressive/self-destructive. There's "America" in a nutshell, regardless of race, creed, class, gender, "sexual identity," or any of the other superficial ego-categories that people hold so precious. Never has so much meant so little, as does now in the legal fiction called "the united states."

The classic gnawing-off-of-one's-own-limb of animals-in-cages is not literally done by most people living "inside," but the internal/"spiritual" self-mutilation is obvious enough if you care to open your eyes to it. Winston Churchill said, "We make our houses, then our houses make us." Indeed. The culture has made millions of toxic houses, and now we have an entire Society of toxic people. To make a long story short, I've become too feral at this point to subject myself to such torture, and I don't believe many people would really want me in their inside space. My mere wild presence would inevitably cause discomfort for all but the most open. I scoff at the idea of "Rules," starting first with my own, thank you very much, which I have gradually been locating and weeding out. The freedom I seek—and already experience—flies in the face of the Social Contract.

I have been increasingly finding myself in perceptual spaces that are beyond words such as "happy" or "contented" or even "blissful." That is, my state of being has been finding itself in non-linguistically expressible places/times that are no longer in "place" or "time" as such things are commonly spoken-of. There is a growing comfort/acceptance with the substantive fact that I—like everyone else, whether they let themselves feel it or not—am *alone* in the world. I do not mean "alone" as in "without human company" or "lonely." Rather, I have been seeing that each of us is moving (or being moved) through the experience of life on-our-own, with ourselves only, in a singular union with everything currently around us in any given moment (that's not a clock-moment). The *immediacy* of living experience is, in fact—not in theory—all there is. The simplicity of it all—of "reality"—is denied and repressed by nearly all aspects of Society, with the result that most everyone acts not from what is in front of them, but from Fear. Fear of nothing at all, as it turns out, except the delusions of the ego/mind. This state of dis-ease is not inevitable, and is not the natural state of living— that's just what we tell ourselves because we are afraid of facing our own wholeness and perfection.

None of these feelings/thoughts are new or "mine." This is basic living-in-a-body stuff, and nothing to get excited about. Born in another place/time—i.e., Africa/12,000 years ago—none of this would even have come up. This is a purely modern conundrum.

In the meantime, we have farmers grateful for a dry space to sleep on this wet night, such as Demeter, who is in the greenhouse, curled up in a chair amongst the bins of wheat, fortunate that it hasn't rained enough for it to be dripping in that particular spot, like it did all last Winter. Need to tar that seam in the roofing when it dries out, if I remember to get to it. The water stain on the inside beam shows where it needs attending....

09.19.2009: "Farmer K' of Sunroot Gardens pranks local men's wellness group"

It was (and is) my strongly held belief that knowledge should not be monetized, especially if it relates to skills essential for survival like food-growing. One of the central practices of Sunroot Gardens had been the free sharing of anything and everything we had learned. I was peeved by people who charged money for gardening or farming workshops. There's nothing new under the sun when it comes to agriculture, so anyone selling something was trafficking in stolen goods as far as I was concerned—stolen by them from the public domain.

In the summer of 2009, I was approached by a local men's group about giving a class on winter cover-cropping as part of an educational gardening series they had organized. I was happy to do so and accepted the invitation without question—or investigation. The day before the class, it came to my attention that the group was charging for their events, including the one I was leading. I was incensed, but did not cancel. Instead, I decided to make my statement when the students arrived. Later that day, I put out a fake press release about what I had done. Besides sending it to the Sunroot email list, I also posted it to the Portland Indymedia website, where the comments expressed complete incomprehension. Looking at the piece now, I can see that I was truly in my own space at that time, uncaring about whether I was understood or not. I don't see anything wrong with that, but it's no surprise that I wasn't making much sense to other people.

Note that the text is carefully composed so as not to officially ascribe a gender to 'Farmer K'. This is a game I have continued to play when the need arises to write about myself.

Date: Sat, 19 Sep 2009 13:37:40 -0700
**Subject: [sunrootgardens] 'Farmer K' of Sunroot Gardens
pranks local men's wellness group**

FOR IMMEDIATE RELEASE

"'Farmer K' of Sunroot Gardens pranks local men's wellness
group"

City of Roses, Cascadia - 19 Sept. 2009 - This morning, shortly after 10am, 'Farmer K' of Sunroot Gardens, an urban survival gardening enterprise, pranked Manifest, a men's wellness group.

Earlier in the month, a member of Manifest had contacted 'Farmer
K' about teaching a class as part of Manifest's Fall/Winter Gardening series. Taking into account the Season and the resources that
Sunroot Gardens had to offer, it was agreed that 'Farmer K' would
play instructor for a class on Cover Cropping for Winter. The class
was to go from 10am to 1pm on Sept. 19, and was to include a
lecture component, a tour of gardens where cover-crops are already growing, and perhaps the planting of new crops in yet other
gardens.

Cover cropping is an agricultural practice with many methodologies (and non-methodologies) that varies from farm-to-farm, re-
gion-to-region, and season-to-season. In Cascadia, mild Winter
temperatures and plentiful precipitation allow for a broad variety of
plant species to be grown as cover-crops. Whether in a field of
acres-in-size or a home plot of just a few hundred square feet, a
winter cover-crop serves the same purposes, among which are:
the crowding out of unwanted, non-agricultural plant species (aka
"weeds") until desired vegetable or field crops are planted; the improvement of soil tilth through the production of organic matter that
is turned into the soil in the Spring (aka "green manure"); the addition of soil nutrients through the mineral-accumulating activities of
particular plant species (such as nitrogen-fixing Legumes); and,
the protection of friable soil tilth from the compacting effects of
months of rain.

Many plants used as cover-crops also provide edible food: i.e.,
Grass Family species such as oats, wheat, and rye; Legume Family species such as field peas and fava beans; and Brassica family
species such as radish. As part of an urban survival gardening effort, cover-cropping is arguably essential.

112

'Farmer K' opened the class by handing out cell number and email address, so that everyone could make contact on their own time, at some later date -- not that day -- for reasons that would shortly become obvious.

"My knowledge about cover-cropping comes partly from books and talking to other farmers, but mostly from my own experiences and observations," claimed 'Farmer K'. "I am delighted to share everything I can, but I have nothing to sell."

'Farmer K' then noted that the fact that the day's class was not free-of-$-charge was not made known until the issuing of Manifest's email announcement thereof, to which 'Farmer K' was cc'd.

"Agricultural knowledge is in 'the Commons' of the human race," 'Farmer K' asserted. "It has no place being commodified. Perhaps there were people who would have come here today if there had been no charge."

'Farmer K' did not mention that on this particular morning, the Sunroot Gardens current floating cash balance would not have been able to provide for the $15 "drop-in" charge, had 'Farmer K' been a person wishing to attend.

'Farmer K' then repeated the willingness to share resources, both informational and material (such as cover-crop seed), to everyone present, but "not today". These words were followed by the informal sign language for "my lips are sealed", of a mimed key locking the mouth and being discarded, and then, after a little bow and entirely sincere smile, departure from the garden by bicycle.

Two Manifest members found 'Farmer K' a few minutes later at the Division Stumptown handing out cherry tomatoes in three colors and loads of Italian prune plums to anyone who wanted them, customer and barista alike. There, one of the Manifest members delivered an apology, which 'Farmer K' smilingly shrugged off. "There's nothing to be sorry for," 'Farmer K' averred, while drawing a puff of homegrown tobacco from a corncob pipe. "I have no hard feelings at all."

Sunroot Gardens will be sowing Winter cover-crops of various types and with various methods (and non-methods) at plots of garden- and acreage-scale for the next two or three moons in the Southeast quadrant of the City of Roses metropolitan area. Folks wishing to learn what they can about Winter cover-cropping are invited to join up with any of the upcoming missions as is convenient to their own time and appealing to their own inclinations. An email listserve announces some activities. One can also just call or email and make an arrangement.

SUNROOT GARDENS is not a business or a non-profit, and is not affiliated with any political group, religion, or philosophy. It is simply an effort to feed as many people as possible by offering ways to participate in the growing of food. Participation with Sunroot Gardens is without set or established terms, conditions, or limitations, though fresh produce and home-cooked meals have often been offered as an expression of basic hospitality. Sunroot Gardens does not provide a "safer space", in which policing of the self and others is demanded; rather, a "wilder space" is offered, in which everyone is invited to explore their own freedom apart from the strictures of 'Society'. This includes freedom from Society's superficial foil, 'Rebellion'. Being 'against' the Establishment gives it as much power in your own life as being 'for' it.

'FARMER K' is a persona manufactured by the imagination of the beholder and has no actual material, concrete or substantive reality. As in, "I know that guy," in which the speaker is referring to an article they saw in the media, or—with another degree of separation—merely heard about. Claims of having met 'Farmer K' deserve equal skepticism, as—again—all that is generally being described is the structure of a persona created in the so-called witness's mind. Pardon the paradox, but, for 'Farmer K', there *is* no 'Farmer K'. :-)

MANIFEST describes itself as "a non-profit men?s wellness community [that] helps Portland-area men pursue their wellness visions & passions together in more than 40 groups & classes monthly like Yoga, Sex & Intimacy, Work & Abundance, Cycling, Walking, Gardening & Food Sustainability, Meditation, Affordable Housing, and Healing Touch Exchange, coaching, and more." For more information, see: www.manifestpdx.org

###

(1069 words)

09.27.2009: "Back from the Forest (but not Out-of-the-Woods)"

"Tell me, who are you, alone, yourself and nameless?" (Tom Bombadil, in J.R.R. Tolkien's "The Fellowship of the Ring")

September of 2009 was an intensely emotional month for me. That spring, I had found myself in a love affair with one of the farm helpers, a young man also originally from the Midwest. Like most love affairs (in my experience, at least), the fun part didn't last long and soon it turned to heartache. The break was acrimonious and happened shortly before the bike-trip described below. Though the relationship was a major element of the 2009 season for me, I have chosen not to mention it anywhere else but here, in this chapter. So, though this book is revealing, it is not technically a "tell-all."

Date: Sun, 27 Sep 2009 21:06:23 -0700
Subject: [sunrootgardens] Back from the Forest (but not Out-of-the-Woods)

On Monday I took off by bicycle for a solo camp-out in the forest south of Wy'East (slave name: "Mt. Hood"). On Saturday, I biked back. I appreciate that the Wednesday and Friday harvests/distro were taken care of by Demeter and by Phil of Mall56.

This time of year is a slight lull in the agricultural season. Winter planting is mostly over, cover-cropping still has to wait, and the fall crops like quinoa, soup beans, & winter squash, are not ready to pick yet. So it was a good time to "get away" for a few days.

Lull or not, I went because I felt a strong desire to do so, and could not imagine resisting the desire. I had another Harvest that needed tending: the one that follows the apparent departure of someone who had been a lover.

If you're looking for Farm News, you can stop reading here, as the remainder of today's message is non-agricultural.

It was Angel who referred to the jumble of feelings and thoughts I was experiencing in the days following the departure as a "harvest." This was the time, he said, when one reaps what was sowed. All the fertilizing, attention, observation, etc., that went into trying to make something flourish were now bearing their fruit/seed. He cautioned against resisting any of the emotions that I was experiencing. None of them were "abnormal"; quite the opposite; in the moment of feeling anything, *that* is who I was, be it sad, angry, relieved, haughty, and on and on. All of these things were doors to deepness, opportunities for understanding. Rejecting any would just hurt me by denying the truth of myself to myself, he said.

I got a ride for me and my laden bike to Estacada with Echo, and we stopped at his farm on the way and sat by the river talking for a good while. I had originally met Echo up in a tree-sit at the location I was aiming to reach by bicycle, so it was an appropriate send-off. He was also able to give me good directions and point out where the place was on the map. Though I had been there maybe 10 times, I had never gone by myself, so wasn't sure of the way.

A head-wind was coming down the Clackamas River valley the whole first day, and I camped the first night at a crossroads called "Three Lynx Junction." The quantity of stars I could see through the thick trees above me far out-numbered what is visible in the entire sky in the city. I fell asleep to the rushing of the water, which eventually won out over my fretting mind.

The second day's travel was probably the single most arduous physical experience I have ever had. It's true that farming, especially by bike, is arduous, but not with the intensity that this day's journey brought.

I had to walk the bicycle for most of the day because the roads were so steep. Not Seattle-steep, but longgggggggg and steep. I estimate that I walked the bike at least 12 miles, and maybe more than 15. The destination was at around 4000 feet in elevation. The hills were so seemingly endless that several times I started to cry when turning a corner and seeing that it was *still* going up. By the last mile I was starting to stumble, and the pain in one of my thighs was becoming overwhelming. There is an old injury on that side that expressed itself in this way.

But being obstinate about it, I wouldn't settle for anything less than the destination I had in mind. Just because I was familiar with the spot, and knew that its remoteness would virtually guarantee me privacy. The City had been feeling claustrophobic and I just wanted distance; the distance to be closer to myself, whatever that might mean.

Late Tuesday afternoon I arrived and collapsed. I found the energy to gather some wood for the stone fire circle I had made last time I was there, in 2007, which was still intact. My sleep came fairly easily and in the morning I was woken by a chill north wind.

I spent the first part of Wednesday picking another camp spot out of the wind, building a new firecircle, gathering wood, and going to a nearby stream for gallons of water. I cooked some good food and ate it.

Thursday was finally a day of no travel and no set-up. There were now no housekeeping duties to distract me from the chatter going on in my head. The chatter had been constant since leaving for the trip. Almost every moment of pedaling, pushing, gathering, etc., had been executed without any real presence with them. That is, here I was doing these things—and now that they were done, just hanging out in a forest of majestic old growth trees in a preter-naturally quiet place—but none of it was quite real. The chatter in my head was that loud, that insistent, that relentless. I looked at the trees and realized that I wasn't even seeing them. The world I was living in was a world of suffering that existed only in my head. This suffering was not present in the physical space of the forest. But the suffering was blocking out my experience of this physical space.

I kept in mind what Angel had said and I did not try to make the suffering go away. I tried just to be present with it, since it was, af-ter all, the entire world that I was living in at that moment.

I smoked my homegrown tobacco and every once in awhile a sob would escape my throat, but no real crying. I saw what looked like a comfortable spot under one of the big old trees—at least 200 years old—and sat there. I closed my eyes and felt that the mass of the tree extended under the ground under me and all around me. It felt like I was being held up by the tree, as if no ground ex-isted at all. I leaned my head against the rough bark and then something happened that I had never experienced before. I found that the tree and I were in conversation.

The sense of a question would come into my head, and before it had formed in my consciousness, the answer would appear along with it. Then the words of both would unfold for me in conscious thought.

One thing the tree said: "For me to say 'I' and for you to say 'I' means the same. (Nothing.)"

This came too, but I am paraphrasing, for the "words" of the tree were experiential, not memory-making: All living things are just Life. There is no individuation except in the mind. Me, the pathetic monkey running around fretful on the ground below these giants—this was my own delusion. In countervailing force to "gravity," Life leaps up with "levity" in the meeting of earth and sky. (Of "mother earth"—with her minerals and elements—and of "father sky"—with his influencing sun, moon and stars—if you'd like to think of it that way.)

Gathering wood a few minutes later, watching the fretting-fretting-fretting that was *still* going on in my mind, I suddenly "saw" something that shocked and overwhelmed me.

I saw that all my mental/emotional suffering—in this moment and in every other that I had experienced in my adult life—had been wildly misplaced. I saw the pain over which all the suffering had been about. It was one single, deep, sharp pain. It was the pain of disconnection from Living itself. The pain of living in imagination, not immediacy. The pain of not being in my body the way the tree was, in union.

Then I saw that "suffering" and "the sufferer" are the same. They come into being together, never one alone. The creator of both is the ego/mind, which seeks to define itself as a separate, special, individual thing somehow apart from the world in which the body lives. This is delusion. It is *dis*union.

Then, all at once, I saw that I had been holding the recently de-parted lover, and the other lovers I have had in my life, as being like precious jewels in my mind-memory. Yes, sadness had ac-companied the partings with all of them, but they were all somehow "significant," something to be treasured: special. I saw all at once that this was ridiculous, romanticized claptrap. I had a vision of the jewels turning into a cheap dime store bracelet, and I was a stupid monkey fingering the plastic beads with a delusional greed.

In fact, I saw, all the sadness of disconnection I had felt with those lovers was a foil, or a shield, put up by the mind to mask the deeper disconnection I had (and have) with Living itself. Yes, the suffering over the lovers was "real," as was "I" as "sufferer." And yes, the experiences of connection that I felt with them in various moments were also "real," as are all experiences of connection with *any* other living thing, be it kale or cat. In those connections perhaps I had moments of true connection to Living. I don't know. The memory is so muddied by the mind and its interpretation and reinterpretations.

If I did feel true connection—union with life, free of the mind—with lovers, then they were only experiential moments. This is an "only" that is enough in a life not disconnected. But, soon enough, the mind would step in and try to connect memories of these moments into a narrative, whether I called it a "relationship" or not. Expectations would enter the picture. Projections into the future or reworkings of the past. "Meaning" (which for the tree is nothing at all). All this in the service of the mind just plain making something up. There was the calling-into-being of the suffering/sufferer.

This glimpse of "The" pain, if you want to call it that, was terrifying. I found my mind imagining a fear: a life in which the suffering/sufferer never goes away. But I also saw in that same moment that "hope" for the suffering/sufferer to go away is no different. Hope and fear are the same, and equally as futile. Both are the frettings of a deluded monkey. Now I remember something U.G. Krishnamurti said: that having freedom from desire includes freedom from the desire to have freedom from desire. Don't try to wrap your head around that one—it doesn't make sense that way.

After all this I had a load of wood gathered and I stood up, looked around at the trees, and said, "OK, now I don't *have* to be here. Now I can be here just because I *want* to."

The next day, Friday, was my last full day there, and I finally felt re-covered enough physically from the ride to walk around some. I re-found a little meadow in one of the old growth clusters, where I had spent time on previous trips. I enjoyed just sitting there, looking at the light coming through the needles, and listening to the birds, one of which had led me there in the first place with its distinctive sound. I visited the tattered remains of another old growth pocket further down the road and marveled at the immense root ball of a 200+ year old tree that had tipped over some time in the last year or so. I took a rock from the heart of this mass as a present for Sister Melanie, the farmer down at Lovina who I farmed with around town by bike last year.

The fretting over the departed lover did not end, but it was more subdued. I had seen that, as stormy as the feelings and thoughts were, there was deeper pain that trumped them. This was strangely reassuring.

THE WIPE-OUT

On Saturday, I started the bike ride back early in the morning with the idea of getting all the way back to Southeast in one day, even though it was 80+ miles. I figured it would be easier since the load on my bike was now much lighter and most of the way would be downhill. But looking at the smoke from the fire, I saw that the wind changed, and I suspected this meant I would have a headwind coming back.

Down the hills I went, grateful that I had worked on my brakes the night before. On the unpaved road I was probably going 20+ mph. On the paved roads, over 30. I had some exhilarating moments, looking out over the vastness of the forest, the ridges bathed in golden light, the steep drop-offs—almost cliff-like in places—on the one side of the road. Here, too, perhaps I had some flashes of unminded connection.

When I got back on 224, along the Clackamas, the main road back out, the wind started. Oh, how I cursed in places where the angle of the valley brought the full force of the headwind on me. On one stretch, the bike would come to a complete stop within 15 feet if I stopped pedaling. And that was going *downhill*!

So now I was fretting not just about the lover, but about the road conditions, and how disappointed I was that the ride wasn't going to be a breeze [couldn't help the pun!]. It felt like a fight getting myself to move.

I stopped, put more air in the tires, and dumped out almost all my water. I was, after all, biking along the Clack and could get more water whenever I wanted. Then, on the next big long stretch down-hill, I crouched down to make myself more streamlined. I finally be-gan to pick up some good speed.

But that custom bike with the buckets does not behave like other bikes, and when I crouched down, I f'd up the center of gravity. It was not apparent at first, but then, rounding a corner and heading down an exposed slope, the bike began fishtailing, pretty wildly, pretty fast. I tried adjustments to the steering, and the brakes to no avail. I tried to straighten up, but then the wind got the best of the situation, I lost control of the bike, and wiped out, right into the lane of traffic next to me.

I have wiped out before and it turns out I have known in those mo-ments how to fall well. I've never hit my head, and this time, I caught myself without the impact getting my elbows. The wipe-out separated me and the bike, and the first thing I did was look to see if anyone was coming from either direction, which they weren't. So I quickly picked up the few things that had fallen out and wheeled the bike over to the other side of the road, by the river, where the shoulder happened to be quite wide. A fisherman came up to check on me, twice, and was reassured when he saw that I had a first aid kit and was taking care of myself. I just went right to work cleaning the bleeding scrapes on my legs, applying tea-tree cream, and then covering them with gauze that I held down with medical tape. I felt the adrenalin surging through me, but no panic, or indeed any emotion. The fretting—about the lover, about the wind—were gone.

I figured I had a couple miles before the physical pain from the im-pact kicked in, and I knew that the market on the edge of the forest was not much further than that. I felt quite calm. The wind didn't seem as strong. The bike's gears, which had been acting up every few hundred feet, suddenly worked nearly flawlessly. And the fret-ting was a tiny distant voice that I could hardly hear.

After a few minutes I started to feel the places on my body that had been impacted, but the market appeared on the left about a half mile later. I got some food and asked the sister working there if they had Advil or anything. They didn't, but when she heard I had wiped out, she offered to get some Motrin from her car. I appreci-ated her generosity, and my body appreciated the 800mg of pain reliever, which made the rest of the ride entirely tolerable.

For the rest of the day, I just pushed myself at the strongest pace I could, taking frequent breaks for food/water. In Estacada I checked out the new health food store run by the 7th Day Adventists, then rode the "old road" up on the ridge above 224, to avoid the fumes. Four or so miles of 224 were unavoidable, but I stopped at a farmhouse and got some fresh eggs. Then I pushed the bike up an enormous hill to Boring (to avoid a worse hill into Damascus/Carver) where I found the end (or the start) of the Springwater Corridor.

As I rode the Springwater Corridor. into town, I noticed the changes in people. The openness that people have in the wild areas was replaced little-by-little by fear. It felt like entering a prison, honestly, as I watched the level of disconnection get stronger and stronger. I felt free to move among it, since it was a prison where everyone was making the bars themselves, in their heads. The alienation—self-imposed—was palpable. In the forest, people at least wave to each other on the backroads, and say hello at little pull-offs. Sometimes much more. There's less judgment based on appearance, and more connection to the senses.

By the time I had gotten into Gresham, people were regularly ignoring a nod of the head or tip of the hat, and in some cases betraying their fear very clearly on their faces. Parents were corralling their children like cattle, and tugging their dogs on leashes. The fakery of it all was very obvious; how most everyone was turning off their senses and living in a world of zombie-like obedience. The people who showed the most sincerity in the interactions I had on the Corridor were the home-bums, for whom "Hey, brother!" actually means something.

So I'm back in town, pleasantly surprised at the tremendous growth spurt put on by the Nicotianas and the Spilanthes and the winter greens. I am not in this moment fret-free, or in union with Living—not "out of the woods" yet—but between talking trees, a shocking glimpse of my deepest pain, and a good bike wipe-out, I feel pleasantly shaken up and settled down, if that makes any sense.

We'll see what happens tomorrow, if there is one. I've got no idea what to expect.

— "Leave life to live itself." U. G. Krishnamurti

10.09.2009: Quinoa Consciousness

I had been experimenting with growing quinoa since 2005, the first season at Lemon Balm Garden, saving the seed each year and replanting it the next. Starting with two packets of the "Faro" variety—purchased from the Wild Garden Seed company in Philomath, Oregon—I had increased my stash to 36 pounds by the end of 2008.

I will always remember my first meal of homegrown quinoa, in the autumn of 2008. After soaking the seeds and rinsing all the bitter-tasting saponins off, I cooked up a pot and ate it unadulterated except for a pinch of salt. Quinoa from the store had never tasted so delicious. As I savored each bite, pictures came into my head of specific Quinoa plants I had known from the field that year. Unlike other staple crops like wheat or corn or buckwheat, each Quinoa plant was unique in appearance. I spent long loving hours with that patch and with particular specimens within it. I had also known their parents and grandparents. Was I so close to this seed and to the creatures who had borne it that I was able to detect and then visualize from memory the individual plant that was contributing to the forkful in my mouth at a given moment? Maybe. The experience *felt* magical, that's for sure.

2009's planting was the largest ever. The weather that year had been perfect, the plants had grown well (though not as well, of course, as we hoped) and rain would not be arriving early to spoil the harvest. I announced harvest time:

Date: Fri, 09 Oct 2009 20:08:06 -0700
Subject: [sunrootgardens] Weekend staple crops work parties

...The sunny dry weather is supposed to hold through the weekend, so we still have a window to harvest and process staple crops. By "staple crops," we mean those foods besides fruits/veggies that provide the majority (85-90%) of the human diet. Grains, beans, oilseeds, etc. Sunroot Gardens has 1 1/2 acres of such planted in Milwaukie.

We put a huge dent in the harvesting and processing of crops at the Hampton acreage this week. On Saturday, we will be returning for QUINOA harvest. Farm folks will be there by 9am, and we will stay as long as people want to stay, or until it's too dark to see. If you have never seen quinoa, this is a great opportunity. We have over 1/4 acre of it planted, and it grows in many colors. Doesn't look like wheat, or corn, or amaranth, or anything else.

If you have never experienced a grouped-plant-consciousness, the quinoa can share that with you, too. Having grown this seed out since 2005, they feel distinctly like "brothers" to me, personally. Through my moments with them, I have come to suspect that plant consciousness is time-spanning in a way that is not linear like human consciousness often tends to be. That is, it has felt to me as if the future generations of this quinoa and the past generations of this quinoa are equally as present with the current generation, and can therefore give one of us a glimpse in either direction.

Be that as it may, quinoa is a high protein grain (nearly a meat replacement) that is drought tolerant and gluten-free, and might be integral to the emerging locally grown Cascadian diet that we are discovering together.

10.15.2009: Set Your Own Prices / The Dollar-a-Year Farmer

For the final season of the CSA—from the autumn of 2009 through the autumn of 2010—I introduced another radical innovation: price-your-own-shares. I have never heard of another CSA pulling a stunt like this.

Date: Thu, 15 Oct 2009 23:12:48 -0700
Subject: [sunrootgardens] Winter squash! / 2009-2010 shares available

...The Farm is not pricing shares this year. Instead, we are presenting the year's budget and inviting people to offer what they would like to offer for what they would like to have. We are putting the pricing into your hands because you are the one who knows the most about how much you'd like to take, of what, how often, etc.

Buying into this year's budget gets you produce from Winter Solstice 2009 to Samhain 2010, which is a little over 10 months. It also gets you staple crops harvested during 2010, to include wheat, corn, quinoa, millet, soup peas, soup beans, etc.

You decide if you want produce or staples or both and then you make an offer accordingly.

...With this new price-your-own-share method, there is nothing stopping you from paying $1 and then picking up produce 3x a week the whole year and not feeling bad about it at all, if that's how your world works. And if that's really how your world works, go for it!

— "Where there is no law, there are no criminals." -Peter Goodchild

I also significantly reduced the farm's budget for that year, continuing the contraction trend started in 2009, and lowering my own salary to $1:

> ...The Farm budget is being reduced this year, by approximately 30%. The Farm is free-er, the less dependent it is on money; in times of economic/social collapse, so-called "underground" economies based in communities naturally emerge that "do business" in different ways. Sunroot is merely riding that wave, and much farm "business" is now done without money. Macro-economically, we are not in a "growth phase" right now; quite the opposite, so the old models of "getting bigger" every year no longer make sense.

> ...Also reduced in this year's budget is the "farmer salary" line item to $1/yr. (You've heard the phrase, dollar-a-year-man?) How will the farmer get by? "Rent" is picked up in the new budget item, "farmer quartering," which is intended to include the renovation of the greenhouse, and the paying of utilities related to that. So for $1800—the monthly price of a condo on Hawthorne—we plan to quarter the farmer for a year. Another line item new to this year's budget is "farm kitchen," meant to help cover for all the food that Mrs. K cooks for the farmhands, including those delicious field meals she has packed up in the past. Hardly a day went by at the Firepit this year when someone was *not* joining Mrs. K for breakfast, second breakfast, elevensies, luncheon, afternoon tea, dinner, or supper. (Or occasionally a late-night-fiend-treat.)

By the end of 2009, I had built enough relationships in Portland that I could make financial choices like this and have it work for me. But it was only possible because Lady Quince, my landlady at the Firepit Garden, was charging me a pittance for utilities and no rent at all. Her generosity single-handedly empowered me to live in the city and pursue the farming full-time. At the time, I fully realized that the arrangement was one-of-a-kind and I knew that it wouldn't last forever, one way or another, so I was very grateful for it.

The fact that Sunroot Gardens and my lifestyle could not be imitated —because its circumstances, though carefully cultivated, were a unique product of timing, my own previously-built networks, and plain old dumb luck—was unappreciated by most people. They wanted me to present them with a "model" they could imitate. My own apparent ease with my way of doing business had the unintended consequence of making the whole thing *look* easy to outsiders, who were—for the most part —entirely unprepared and unwilling to give up the things I had given up to attain my enjoyment: a real house, a conventional social life, "time off." *I* did not miss these things, but they were exactly the things that most people were actively seeking and would not find if they used me and Sunroot Gardens as a "model."

Though I spoke and wrote about these conditions constantly (or so it seemed to me), the point was repeatedly missed. For me, two things made my choices worthwhile: my own personal freedom and my intimate plant relationships. But by the end of 2009, I felt like the city was limiting both. Only my closest friends saw the reality of my situation for what it was and understood that it was not sustainable for me in the long-term.

In-Depth — Beyond Salads: All about the Staple Crops Project

References to the Staple Crops Project are sprinkled liberally throughout this book, but the best summation of our efforts as a whole are found in this report, which spans the 2008 and 2009 seasons. Having spent some time in the business world when I was younger (and having been raised by a father who was a Professor of Economics), I had an appreciation for data and spreadsheets as tools for measuring investments and outcomes, so I applied this method of analysis to our Staple Crops efforts. I had a genuine curiosity about whether the project was "penciling out" but I also wanted to explicate the challenges involved with hard numbers. Too often, the "sustainability" crowd in Portland was (and still is) dreamy and untethered and I wanted to give them a healthy serving of reality so that they would get serious.

STAPLE CROPS REPORT 2008/09

I. Introduction

II. Financial & organizational model

III. The Plots

IV. Methodologies & Yields by Crop, 2008

V. Methodologies & Yields by Crop, 2009

VI. Conclusions

VII. Notes

VIII. Photos

I. Introduction

Sunroot Gardens is an urban agricultural effort based in the South-east quadrant of the City of Roses, Cascadia, under the gaze of Wy'east. Sunroot Gardens has operated as a CSA (Community Supported Agriculture) enterprise since 2007, utilizing back/front/side-yards, empty lots, and public rights-of-way for growing vegetables, fruits, and medicinal herbs. Famously, this network of gardens was discovered and developed by bicycle, and bikes have remained at the heart of travel and transport for Sunroot.

In 2008 and 2009, Sunroot attempted to grow various grains, pulses and oilseeds under the auspices of "the Staple Crops Project." While the goal in both seasons was to grow enough food to feed some number of people for some portion of the year, it ended up functioning merely as a research project. Yields were consistently lower than hoped for, and the logistics of harvesting and processing proved more difficult than suspected. Over $15,000 was spent, much of it out-of-pocket for the farmers. The Project has been extended into a third year, 2010. We have included our projections and plans for 2010, which has entirely different logistics and circumstances, in the "Conclusions" section of this report.

This report from the Staple Crops Project is intended to give an unvarnished, unsentimental, just-the- facts-ma'am view of our experiences over the last two seasons. **The Staple Crops Project is not intended as a "model."** There are no "models"—there is just life, in front of us, and there is nothing to do but live it, whether we'd like to admit this or not. We are not interested in entertaining the intellectual and emotional illusions that permeate the whole of the "Sustainability" movement, so herein we offer these facts, figures, and observations only as a record of what we saw in front of us.

II. Financial & organizational model

Sunroot Gardens was founded by Farmer K in 2007 as a CSA farm, an arrangement in which people pay $ to the farm up-front, and receive produce throughout the following season. Recognizing that produce—as in vegetables and fruit—comprises less than a quarter of the typical contemporaneous human diet, and that the majority is made up of grains, pulses, and oilseeds/nuts (with the addition of animal protein for many people), Farmer K saw a need to expand Sunroot's efforts. Thus the formation of the Staple Crops Project was announced in 2008. A budget was created:

```
LINE ITEM............................AMT, US$
Land-lease ($100/acre/year)...........200.00
Soil Test.............................33.00
Seeds.................................493.61
Cover Crop Seed.......................280.56
Amendments............................600.00
Tools.................................397.60
Harvest / Processing..................544.95
Incidentals including fuel and
LIWHWYMOP (Life Is What Happens When
You're Making Other Plans)
aka unforeseen costs..................350.00
TOTAL:................................2549.72
```

Ten shares, at the cost of $250.00 each (~10% of the total), were made available. All were sold within two months of the Project's announcement. Recognizing that labor is as important as $ with an agricultural project, the final harvests for 2008 were divided as such:

* 40% to the $-shares (with each share receiving 10% of that 1/3)

* 40% to the helpers (divided in proportion to number of hours worked)

* 20% to the farm (for the farmers' tables, and for seed for the following year)

Actual costs exceeded the total by over $5000, due mostly to the purchase of a tractor:

```
Tractor (1958 35hp Massey-Ferguson):.......$3500
Tractor repair:............................$600
Tractor upkeep:............................$500
Additional amendments:.....................$250
Additional cover-crop seed:................$200
Additional LIWHWYMOP (est.):...............$800
```

These costs were covered by Farmer D, Farmer K's partner farmer in the Staple Crops Project that year. For 2009, a larger budget was created:

```
Land-lease ($100/acre/year)............$200.00
Seeds..................................$560.00
Cover Crop.............................$641.10
Amendments.............................$1000.00
Tractor upkeep, repair, etc...........$1000.00
Harvesting / Processing................$375.00
Incidentals (incl. LIWHWYMOP)..........$425.00
TOTAL:................................$4201.10
```

Ten shares were offered at the price of $420.11 each.

Because the 2008 yields were so low, $-investors from that year had their shares extended into 2009. The divvying up of final harvests was calculated as follows:

* 15% to the 2008 $-investors

* 35% to the 2009 $-investors

* 25% to the helpers

* 25% to the farmers

Actual costs for the 2009 season exceeded those budgeted by at least $3000. These costs were covered by personal investments from two of that year's four farmers. The over-runs were for additional amendments and seed, which was no surprise, but also for equipment costs related to the wheat harvest, which were unforeseen.

For 2010, the budgets of the produce CSA and the Staple Crops Project were combined, so the Staples costs were to be some portion of the entire 2010 Sunroot Gardens budget, which is $10,000. Those who wished to invest in the 2010 Staple Crops Project named their own amounts, and the investor portion of the harvest will be distributed proportionally among them. Additionally, a five-figure chunk of cash has been made available to the Project for 2010 by an anonymous "angel" investor who is not seeking anything in return for their investment. The disbursement formula for 2010 is undecided but we hope to offer shares to the community $-investors that exceed previous shares by factors of at least 50-100.

III. The Plots

In 2008 and 2009, the Staple Crops Project was grown on two main plots: "Carver" and "Bailey's" (formerly "Hampton"). Unless otherwise noted, the various crops were dry-farmed, with no irrigation except rain and field moisture. We generally used the crop-spacing recommendations for dry-farming made by Steve Solomon in "Growing Vegetables West of the Cascades." In some instances, we found we could crowd plants more than he suggested.

The main soil amendment used was AZOMITE, given its relative low monetary cost for the perceived benefits conferred. Bailey's was also limed lightly in 2009, and the corn and some beans there received compost tea. The compost tea was applied in two ways: foliar feeding with a hand-held sprayer and root-zone irrigation with hand-drilled holes next to individual plants.

CARVER

Carver was two acres outside Carver, which is between Oregon City and Damascus, just outside of the Portland-metro Urban Growth Boundary. The land itself was alluvial bottom-land, tucked into the north-then-south bend of the Clackamas River shortly before it spills into the Willamette. The area had been conventionally farmed by another lessee-farmer from time out of mind, and the effects of his heavy machinery and chemical-dependent methodologies were obvious.

The soil was a lifeless, airless clay that made mad muck in the wet winter and dried to a concrete-like consistency in the summer. No worms were to be found. Adjacent plots covered in blooming mustard were the home to no insects that could be seen. A hammer and chisel were needed to extract a soil sample in 2009, and the chisel broke.

(The farmer later claimed to be switching to USDA National Organic Program [NOP] standards, at least for those parcels close to the residents' house; however, doubts were cast on his following of these standards due to: A. his choice of certifying agency; rather than Oregon Tilth [considered stringent], he went with Washington State [where he "has friends"]; B. the spraying of "something" on more than one occasion; C. the different appearance of the farm on the day of the inspection. NOP does not disallow the use of tilth-destroying large machinery, which continued as well.)

The $100/acre/year arrangement for two acres was initiated by the residents of the farm (who did not farm it themselves) who were wishing to push back the conventional farmer from their house so his spraying wouldn't be so close. It was contractually agreed that Sunroot would be using "sustainable" methods that would not harm the residents.

Going into 2008, the two acres were already seeded with an over-wintering wheat crop which Sunroot chose to leave and harvest that summer. An early attempt to till some of it under with rototillers (after hoes proved to be grossly adequate) and seed soup peas led to discoveries about the unworkability of the soil, at least with such unburly equipment and at that time of year. Just letting the wheat go seemed the path of least resistance. Going into 2009, we were given two different acres, again already seeded with wheat. AZOMITE was the only soil amendment used at Carver in both years.

BAILEY'S (formerly HAMPTON)

Hampton was 1 3⁄4 acres in Milwaukie (the next town south from Portland within the same urban contiguity) that was owned by the brothers who ran the convalescent home next door. In 2008, arrangements had been made by another pair of farmers to farm the land, and Sunroot was invited to share the space since they wouldn't be using all of it. For 2009, Sunroot took over the entire piece. The land was south-facing and uphill from the railroad tracks, and had never been built up. Horses had—perhaps—been pastured on a portion of it; otherwise it had been fallow since for-ever, just getting mowed by the brothers. The vegetative life there was mostly grass and Queen Anne's Lace and other regional pio-neers, with some thistle and blackberry. The tilth was quite decent. A creek bed followed one side of the field, running in 2008 but dry in 2009.

"Hampton" was named after "Mr. Hampton," a cat who lived with the couple who brought us to the plot. "Bailey's" was named after the dog next door, who would often bark at us to play. One day in 2009 he got to come run around in the field, and that's when we found out what he was called, and were able to more properly name the plot.

(As often as possible, Sunroot plots are named for the nearest [preferably on-site] cat, who in the case of "Hampton," was for a creature who never even visited the plot. Bailey, being the on-site animal, took precedence once his name was known. In the Official Sunroot Gardens Nomenclature Stylebook, an "'s" signifies that the creature for whom the plot is named is a dog rather than a cat. "The Firepit Cat" is a special case, in which the people-name for the oft on-site cat is unknown, so the cat has been named for the garden, "The Firepit." This morning, she killed two more field mice in the greenhouse storage area where Staple Foods are being stored.)

OTHER PLOTS

Other plots also played host to Staple Crops Project crops, mostly in the form of compact urban plantings for seed grow-outs for later field sowings. These will be mentioned on a crop-by-crop basis.

IV. Methodologies & Yields by Crop, 2008

In 2008, the Project's food crops were:

* Wheat

* Quinoa

* Cannellini soup beans

* Oilseed sunflowers

* plus seed crops of soup beans and flour corn

WHEAT

At Carver, we had 1/3 acre of wheat available to us. The remaining of the two acres was fallow or under peas. Over the course of two weeks, we harvested and processed over 600 lbs. of wheat, using hand-methods entirely, with tools and equipment no more complex than tarps, buckets, fans, baskets, and racks. Meticulous record-keeping that year yielded the following summary of info:

```
Total lbs:.......................636
minus COB:*......................22 lbs.
To disburse:.....................614 lbs.
```

```
DISBURSEMENT:
Farmer share (20%):...............122.80 lbs.

$ Investor shares (40%):.........245.60 lbs.
1 (one) 40% share:...............98.24 lbs.
6 (six) 10% shares:..............24.56 lbs. ea.

Work-trade Shares (40%):.........245.60 lbs.
for gen. Lbr.**..................40.00 lbs.
To wheat workers:................205.60 lbs.
lbs./hr. to Work-traders.........1.00 lb.

HOURS SPENT:
All wheat work:..................245.5 hrs
Work-trader time:***.............205.5 hrs
Harvesting:......................142 hrs
Threshing (est.):................46 hrs
Winnowing (est.):................50 hrs

PRODUCTIVITY:
lbs./hr.:........................2.59
Total # people:..................42
avg (mean) shift length:.........3.50
```

*"Cost of doing business" = wheat disbursed to work-traders during harvest activities in the form of berries, flour and bread

**"General Labor" = non-wheat Staple Crops Project work-trade hours

***Total time minus the Farmer's time

QUINOA

At Bailey's (then known as "Hampton"), we planted ten 300-ft rows of quinoa. The seed was our own, having been grown-out and saved since 2005. Originally from one Seeds of Change packet of "Faro," planted at Old Lemon Balm Garden (up on 39th there, just south of Steele. You've seen that plot. Everybody has. As of this writing, a Russian brother named Alexander has been tending it.). The seed was grown out there again in 2006, as well as at the Cora Garden, which produced the seed for Hampton in 2008.

2007 was a skipped year, so from gram-packet to 36 lbs. was just three seasons.

The plants were big beautiful brothers, running a rainbow of flower, stem and leaf colors, making for a psychedelic display. We used about 5 lbs. of seed. Our final harvest was 36 lbs. A summary follows:

```
Task..............................Hours
Prep and Planting.................56.50
Thinning..........................12.00
Farmer Inspection, over season.....20.00
Harvesting........................16.25
Threshing.........................24.75
Winnowing.........................14.00
Total:...........................143.50
```

This is a productivity rate of 1/4 lb. of final harvest per each hour of work.

Like the wheat, this crop was processed entirely by hand. We found that the quinoa heads, once cut (and delivered to the Firepit from Hampton by bicycle caravan), needed three days to dry in the sun before they would thresh easily. This was during a hot dry spell and the heads were stacked pointy-top- down in tepees on jute coffee-bags laid over racks in the sun. The bags were there to catch the seeds, since some were falling off already.

No attempt to remove the saponins were made, since the only method we knew of was to soak and wash the seeds, and how would we get them dry from that, in such a large quantity? Instead, receivers of the seed were instructed to clean them before consumption. We suspected that these bitter-tasting saponins, which discourage predation by birds and insects in the field, would also protect from rodents and bugs in storage. So far, no reports of moth-eaten or rat-raided quinoa have been reported. The overall harvest from the quinoa was so small that the Farm reserved a larger-than-20% share of it to have a good supply of seed for the next year. The $-subscribers all received one pound per share.

CANNELLINI SOUP BEANS

At Mall56 we seeded a few pounds of Cannellini soup beans. We had ordered from Italy, thinking: if we have to live on beans, they might as well be gourmet. Cannellini has often been called "the best" soup bean. We planted rows about 18 inches apart, with plants thinned to 8-10 inches apart within a row. We used a legume inoculant from Johnny's Seeds.

The 30'x70' plot had been lawn grass until then. We tilled it under with a rototiller on May 7th. (Sunroot has kept on file some fun cellphone video of Melanie Plies, of Backyard Booty CSA, turning the 350+lb. Troybilt around on the tight passes.) We broadcast buckwheat seed out after the first of three passes. Rains brought up the buckwheat, which choked back the grass. We seeded the beans on June 27th, after tilling in most the buckwheat. We harvested the majority of the beans on Sept. 28th, and threshed and winnowed them on Sept. 30th. Final harvest was 30 lbs. A summary of work:

```
Task (by hand, unless otherwise noted)....Hours
Field prep (with a rototiller)...........8
Planting.................................6
Thinning/Weeding/Watering (est.)........12
Harvesting...............................9
Threshing/Winnowing....................2.5
Total:................................37.5
```

This is a productivity rate of about 12 oz. of dried beans per each hour of work.

OILSEED SUNFLOWERS

An eighth of an acre were planted at Hampton. Those within 100' of the creek were eaten by nutria. Those planted further away were seeded too late into a particularly infertile patch and yielded what could be accurately described as "squat."

OTHER CROPS IN 2008

"WTO Corn" at the Firepit Garden. From the three ears that were produced by 18 seeds planted at Old Lemon Balm in 2006. Those 18 seeds were from another local gardener, saved the previous-or-so season. That person got their original seeds from a Mexican farmer during the 2003 WTO Protests in Cancun, where a Korean farmer famously stabbed himself to death atop the security fence around the summit to express the plight of traditional small farmers under pro-corporate policies imposed by the WTO. Final harvest: a couple hundred ears, all kept by the farm for seed for 2009.

"Black Valentine," "Pink Floyd," "Tiger Eye," "Yin-Yang" and other soup beans, at various urban plots around Southeast. "Black Valentine" has been grown out and saved in varying amounts every year since 2005. Final harvests ranged from 5-20 lbs. per variety.

V. Methodologies & Yields by Crop, 2009

KEY: Crop (Location) — Amount planted : Amount harvested

Wheat (Carver) — 1 1/3 acres : 700 lbs

Quinoa (Bailey's) — 1/4 acre : 50 lbs (est.)

Taylor's Hort soup beans (Bailey's) — six 300' rows : 90 lbs

Flax (Bailey's) — 30'x30' broadcast : 8 1/2 lbs

Dry Corn (Bailey's) — 1/4 acre : hundreds of ears

Millet (Bailey's) — 20'x20' broadcast : 29 1/2 lbs

Buckwheat (Bailey's) — 1/3+ acre : 105 lbs

Soldier soup beans (Bailey's) — eight 60' rows : 22 lbs

Hidatsu soup beans (Eel Skin Madeline's) — 3 sisters patches : near failure

Popcorn (Eel Skin Madeline's) — 3 sisters patches : 6 lbs :

Black Valentine soup beans (Sewell) — 25'x25' broadcast : small

Cannellini soup beans (Sewell) — 25'x25' broadcast : small

Urid dal bean (Lovina) — two 30' rows : lost

Urid dal bean (Bailey's) — one 40' row : decent seed crop

Kenyan beans (Lovina) — two 30' rows : failure

Kenyan beans (Bailey's) — one 50' row : 1 plant made beans

Pink Floyd soup beans (110th/112th) — 25'x75' broadcast : eaten by deer

Quinoa (Carver) — 30'x30' broadcast : failure

Buckwheat (Carver) — 40'x75' broadcast : six plants grew up

Oats, hulless (Carver) — 125'x75' broadcast : a few breakfasts

Fava beans (Carver) — 40'x75' broadcast : stunted

Flax (Carver) — 60'x75' broadcast : 5 lbs :

Cannellini soup beans (Echo's Farm) — 1/4 acre, rows : beaten to punch by rain

Amaranth (Bailey's) — 10'x40' broadcast : not harvested

Details on selected crops follow. Inquire with Sunroot Gardens for more information.

WHEAT

The wheat harvest was, in Farmer D's words, "**a clusterfuck**." 2008's two week from-beginning-to-end process was replaced with a six week ordeal of wasted time, too many car trips, poorly thought-out methods, a distracting tool-fetish 5 , and a lack of both communication and cohesion. The resulting processed harvest of merely 700 lbs. (compared to 2008's 636 lbs. on 1/4 of the same area) was, honestly, pathetic.

I, Farmer K, felt daunted by the task of harvesting and processing what was a potential harvest of over two tons, and so delegated the "pack leader" role to someone else. This was an amazing opportunity for him, as he was very new to farming, and such a large project can provide a wealth of practical experiences in a short amount of time. Certainly, I doubt if any local farm but Sunroot would have given so much responsibility to a "rookie," but then again, what were any of us?

The project began, and in some ways ended, with scythes. $250 was invested in ordering scythes from an Austrian company that had been making them since the 1700's. Apparently, no one in the U.S. makes good scythes anymore, which is no surprise at all considering the general dearth of equipment and tools for small-scale grain growing. (More on that subject later.)

The scythes worked fairly well for bringing down a lot of wheat in a relatively short amount of time. However, what this left us with was heads on long stalks. The previous year we had hand-harvested just the heads, stomped them on tarps, winnowed them with fans, and called it done. The addition of the stalks to the process was a complication. In an attempt to deal with the huge windrow of cut wheat that collected in the field—which we couldn't process by our previous methods—two machines were brought in. The first was a thresher made by modifying a chipper. The second was a win-nower fashioned from a bunch of stove-piping and a leaf blower. Both machines ran (loudly) on gas-powered engines. The thresher worked okay when fed just the right way, but broke a large per-centage (about 1⁄4) of the wheat berries. This damage leads to their fairly immediate nutritional degradation, and leaves them more vulnerable in storage. Hundreds of dollars and many many hours were spent trying to set up and make these machines work.

Reflecting on the process later, I saw two points where I could have intervened: first, calling off the scythes would've left more to pick by hand in the field, giving us a form we knew what to do with. (This was the fetish—with the scythes; yes, people "look cool" when they use them, but that's not enough reason to keep using them when they are creating a bottleneck.)

Secondly, later came a day when I did a timed test with a hand-harvesting and processing method. I used my chamomile rake (which we had been using to harvest flax) to remove spikelets (the wheat heads), which went about 10x faster than removing them by hand the year before. I collected the spikelets in a yellow plastic recycling bin, where I threshed them by stomping on them. Then, up at the house, I set up a fan on a chair and winnowed bucket-to-bucket. In fifteen minutes total (harvest- winnow), I had 7 lbs. 11 1⁄2 oz., a rate of about 30 lbs. per hour. I had originally envisioned the making of a "Gallic Reaper" for spikelet removal, which is a 2000 year old piece of Roman technology. The chamomile rake was a miniature Gallic Reaper in design, and did indeed speed up the harvest considerably.

At that point, I had the feeling that I should remove the rookie from the project entirely, but I ended up letting it go by. Weeks later I saw that indeed this had been a vital juncture in the season's tim-ing, at which a bigger harvest could conceivably have been gained had I taken control of the project and steered it firmly. Then again, who knows? All such speculation has limited (if any) utility. To complete the clusterfuck, a volunteer stole 150 lbs. of wheat, so we had to rejigger the distribution amounts to assure investors and helpers got their due. Theft is likely to be an ongoing issue as socio-economic structures break down. Such incidents will be handled on a case-by-case basis depending on circumstances.

QUINOA

Feeling that the 2008 quinoa planting was too sparse, we went the other way in 2009 and thickly broadcast a 300'x50' section at Bailey's. The planting was about two weeks on the late side (taking place near the end of May), the field didn't get thinned enough, and the heavy cold rains came about two weeks on the early side. The result: a preliminary harvest of 50 lbs. of the earliest maturing plants were all we got. The rest was ravaged by the black mold! The quinoa grow-outs had always obviously been a collection of varieties, as shown by the different colors of the plants (though not of the seeds, which are white). I estimate that there were at least five distinct varieties, and that we got the seed of just one of them —the earliest maturing of all of them. We shall go with this seed, then, if we plant more in 2010. Having started as "Faro" from Seeds of Change, this localized selection has been named "Bailey's."

TAYLOR'S HORTICULTURAL BEANS

We planted these in long rows on the late side—must've been early July—but got a great harvest in September. The compost tea foliar feeding they got might have helped, though we had no "control" row. They were quick, productive, and fairly easy to thresh and winnow. Much of the processing was done on tarps right there in the field, the idea being to avoid moving large volumes of things around. (The car fatigue that both Farmer D and Farmer K felt after the wheat clusterfuck was weighing heavy at that time.) The final harvest of 90 lbs. was quite decent, we felt.

```
Total harvest (lbs.).......90
-Seed for farm (lbs.)......25
Total to distro...........65
Farmer share..............16.25 lbs.
Helper share..............16.25 lbs.
$-investor '08 (ea.)........0.98 lbs.
$-investor '09 (ea.)........2.28 lbs.
```

FLAX

Farmer M helped broadcast the flax seed. We had two varieties we let mix up: Brown Flax and Omega Flax, both from Horizon Herbs. We sowed them in late May. They quickly sprouted and took over, not leaving room for weeds (except a few blackberry brambles). We harvested three ways: with the chamomile rake, by hand, and by cutting them down. We threshed by stomping on tarps and winnowed with buckets and fans. The flax on stalks proved the most difficult to process, but made a fine "brown" component in a compost pile.

The biggest challenge with the flax was that a great number of seed heads were still green and immature while a great number were rattling—dried and ready-to-go—even starting to shatter and dump their seed. I found that the chamomile rake would collect the green ones up close to the tines, where they could easily be picked or brushed off by hand, but this of course wasted them. Some green heads also got mixed with the dry ones anyway, and though they mostly separated in the threshing/winnowing, their presence added more labor. Immature flax seeds can be toxic, depending on how young they are, so you don't want them in the final product in a large amount (if at all). Perhaps a better approach is to wait longer for fewer green heads, even at the loss of the earliest-to- mature. It is also possible that a different harvesting method could be employed, since the drier heads tend to be slightly higher on the plant.

Our intended crop was the seed, for its Omega fatty acids and other health benefits. Growing flax for fiber entails an earlier harvest of the stems, and apparently the processing is laborious.

```
Total harvest (lbs.)..........8.5
Total harvest (oz.).........136
-Seed for farm (oz.).........56
Total to distro..............80 oz.
Farmer share.................20 oz.
Helper share.................20 oz.
$-investor '08 (ea.)..........1.2 oz.
$-investor '09 (ea.)..........2.8 oz.
```

THE DRY CORN

Farmer M joined the Sunroot Gardens Staple Crops Project as "Farmer" for the corn crop. He brought "Mandan's Bride" corn seeds with him, which he had saved from a previous planting. Sunroot had saved seed from three other varieties, Oaxacan Green Dent, Earthtone Dent, and the in-house "WTO Mexican" flour corn (see 2008 "OTHER CROPS" for more about that seed). Having heard that decent genetic diversity in corn seed will not result from a crop of less than 200 plants, and being that all four corn varieties had started from fewer plants than that, Farmers M & K decided to plant all four together and let them mix up. The idea was that all four varieties could be reinvigorated this way, and reselected the following year for what could eventually settle down into a new variety.

We planted about 1/4 acre at Bailey's altogether. In addition to AZOMITE, we sprinkled fishmeal along our intended rows. We seed the area using "Seed Sticks," from Johnny's Seeds. They are New England made tools that allow easy seeding without bending over. On a hollow stick about the size of a broom handle (which we guess they originally used when inventing this thing), is a hopper you fill with seeds. When you poke the ground with the stick, a door opens on the bottom and deposits a seed from the hopper. Ka-chink, ka-chink, ka-chink! just like that you have seeds in the ground. A great tool that we highly recommend. We sowed 5-8 seeds into each clump. The clumps were five-six feet away from each other in all directions, making a giant grid. Each clump had at least two and sometimes all four varieties. They received foliar feeding of compost tea at least once. We did not thin the clumps, thinking that, as members of the grass family, they would be fine that way; even happy. This choice also meant not having to select what to leave or take, which could've been tricky since multiple varieties were involved. By season's end, the tillers coming out from adjacent clumps were making contact at their tasseled crowns.

We planted the corn in two different plots at Bailey's, one right next to the upper part of the creek, and one across the field. The patch next to the creek grew approximately twice as productively, apparently due to the higher field moisture. I do also remember that corner of the field having a different mix of "weeds" before we tilled it, too, so perhaps the soil had more nutrition in it as well.

The tallest creek-side corn plants were 9 feet high! These were the Oaxacan Green. The shortest plants—usually the Mandan's Bride —were closer to 5 feet high. The creek-side patch was so vibrant and thick that it was a full-on corn maze by the end of the season. You could feel totally lost in the middle of it, surrounded by the lustrous singing foliage. We harvested ears when their husks were mostly or nearly entirely browned and dried out. We husked the cobs and laid them out on screens to dry further. Some ears were mixed colors of different varieties. The Oaxacan Green was mostly pure for the simple fact that it was the tallest plant, and corn pollinates by dropping pollen from the tassels on top onto the silks of the cobs below. Some Oaxacan Green cobs were above the tassels of other plants. We did not weigh the corn harvest at any point. Shareholders rec'd 10-14 ears each, with about half the crop held back for seed. We have enough seed to plant corn in plots of 1/4—1 acre in 2010.

BUCKWHEAT

We planted a large square of buckwheat nearly a quarter acre in size at Bailey's, plus a 3'-6' wide border of it around much of the rest of the plot. This border was there to signify the planting area, since the brothers from the convalescent home mowed around the edge and we wanted a visual signifier that "something is going on here," to avoid crops being chopped down (as had happened with the previous winter's cover-crop). The buckwheat was pulled by hand, stacked seed-side-up in shocks, and left in the field to dry. Weeks later, we returned with a tarp and were able to thresh most of the seed from their stalks by banging bundles on the tarps. The stalks were left in the field for organic matter, and to re-seed buckwheat from the seeds still remaining on them.

The threshed but unwinnowed buckwheat crop was stored in two garbage cans in a condo owned by one of Sunroot's supporters, and then finished (but not de-hulled) on tarps in February, during the dry spell. We have not found a way to remove the hulls from the seeds. We understand that buckwheat flour can be ground with the hulls intact, and that, alternately, kasha can be made by soaking them off.

```
Total harvest (lbs.).........105
-Seed for farm (lbs.)..........0
Total to distro.............105
Farmer share..................26.25 lbs.
Helper share..................26.25 lbs.
$-investor '08 (ea.)...........1.79 lbs.
$-investor '09 (ea.)...........3.47 lbs.
```

MILLET

In some ways, this was the stand-out crop of the 2009 Staple Crops Project. It was super easy to grow. A hand broadcast of seed resulted in a thick stand of plants that kept the weeds out on their own. The apparent harvest per area exceeded all the other crops. Like the buckwheat, we are not sure how to remove the hulls, and have not yet experimented with methods of cooking the seed without doing so. Millet is late spring/summer-sown and is mature within three months. We plan a bigger planting of millet in the 2010 year, especially since the quinoa—another gluten-free grain—proved so sketchy in 2009.

```
Total harvest (lbs.).........29.5
-Seed for farm (lbs.)........10.5
Total to distro.............19
Farmer share..................0 lbs.
Helper share..................5.7 lbs.
$-investor '08 (ea.)..........0 lbs.
$-investor '09 (ea.)..........1.33 lbs.
```

VI. Conclusions

For Sunroot Gardens, **the biggest challenges of farming staple crops were the logistics of harvesting and processing**. We would like to point out that Sunroot owns none of the land used to farm, and paid $ out for a lease on only one parcel (at the rate of $100/acre per season, for a grand total of $200). **This points out the lie in the idea that one can't farm without owning land, which we hear as excuse quite often.** (Leasing and sharecropping are common arrangements in agriculture around the world; to those would-be farmers currently living in the city, I say to you: If you want to farm, then farm. The premium put on "owning" is neither necessary nor even realistic. And in our current late-Imperial context—with accelerating economic collapse and impending food crises—it is much more important for us to learn how to grow food, wherever and however that happens to be, than to continue to live by illusion. We cannot eat our ideas.)

After two years of trying [and three previous years of practicing], we have seen how woefully inadequate "our best" has been. Even the most impressive community effort (the 2008 wheat harvest) netted only about 1/3 of the total potential harvest. Until the time comes when Everyone In The Village Drops What They're Doing To Help With Harvest, **hand-methodology will not be effective** for processing grains, pulses, and other staple crops at the scale we need to provide food for ourselves. We have concluded that, **for processing staple crops at the small-scale level (1-10 acres per crop), we will need machines to help us**.

The machines we are looking at could greatly improve the speed and efficiency of harvest. We are investigating machines that reap & bind, thresh & winnow, dehull, and bag. One thresher model can process 1000 lbs. per hour. An Italian-made reaper/binder can cut a field at the rate of one acre in two hours. Compare this with the rates by hand and you will see there is a world of difference.

We have also discovered that small-scale grain-raising and the machines to do it with are nearly nonexistent in the U.S. U.S. farm equipment companies basically stopped making machines for this scale in the 1970's, instead focusing on the large farms (as in tens or even hundreds of thousands of acres). There is also not much in the way of old equipment sitting around on farms that we could refurbish. That which remains is mostly rusted out, no longer understood, and forgotten. **Nationwide searches of used small-scale farm equipment for sale have yielded almost no results at all.** Not that this particular technological road dead-ended everywhere; quite the opposite: in Europe, Asia and India, small-scale machinery has continued to be developed and built. **For what we would like to do here, it appears we will have to order equipment from abroad.** Quite an irony, isn't it? Here we are in what is touted by some as the technological pinnacle of the globe, and we cannot find the tools to feed ourselves.

After much research, it became clear that we would need at least $30K to outfit ourselves for the Staple Crops Project for

2010. This is quite an addition to a budget of $10K total, and is outside the scope of the $100-$400 share price structure we have been using so far to raise funds. In response, an anonymous donor with no expectation of return has made a five-figure amount of cash available to the 2010 farm effort to cover the expenses of purchasing, shipping, refitting, running, and maintaining such farm machinery. These resources have the potential to increase this year's harvest to thousands and thousands of pounds, since we would be able to handle 10-20 acres altogether. If so, then **the shareholders who have been supporting the project so far for so little material return will get to enjoy a jackpot**. These machines could also be made available to other farmers in this and future seasons. In fact, the knowledge of their availability could serve as the impetus for more people to go ahead and plant.

"The future's here, we are it, we're on our own" - Bob Weir and John Barlow, 1982

Submitted by Farmer K

13 April 2010 [rev. 2]

This report was publicly presented at a special dinner celebrating the Project and its harvests. Here is an account of that event.

Date: Fri, 12 Mar 2010 11:30:52 -0800
Subject: [sunrootgardens] Equinox Party+Fire / Staple Crops final distro / Peter Salem award

"And the Winner is..."

A couple weeks ago, we held a Staple Crops Potluck. The food was very very good. Included were: cornbread, hominy and tortillas from the dry corn; 2-3 bean dishes; yogurt with millet; the best crackers ever, from the Carver wheat and flax; fresh-baked bread; vegan "chicken" nuggets from wheat gluten and flax; and more.

The party was held at OG ("Original Gangster") Urban Farm. We had a fire. The first draft of the 2009/10 Staple Crops report was delivered. And, the winner was announced for the 2010 Peter Salem Award for Excellence in a Seasonal Volunteer.

Nominees for the award included:

* Peter Salem, whose excellence as a seasonal volunteer inspired the award. Peter is a student at Willamette University in Salem, and rec'd a grant to spend the summer in Portland, where he worked for Sunroot Gardens and for Zenger Farms (120th & Foster). Peter chose to focus on Sunroot for his final report. Peter contributed to the farm's collective knowledge by taking up the study and practice of composting, something I'd never found the time to do, so we got to learn from him. The results of his research and labor have been clear this spring, as I have been spreading the finished compost everywhere.

* X, "the Wheat Stealer." For sheer number of hours worked and amount of personal investment made—more than anyone besides Angel and Farmer K—X was nominated for the award. He, too, increased the Farm's collective knowledge, specifically in the area of compost-tea making and application. X was disqualified from winning, however, for stealing 150 lbs. of wheat from the farm when he departed.

* River, for making his truck available as much as possible as "a community resource" (his words) to the farm. That fine diesel Ford performed countless tasks too heavy or awkward for current bicycle equipment. River also excelled at scavenging and delivering valuables for the farm, including windows, lumber, and construction materials for cold frames and the greenhouse renovation.

* Mike Mandan, who helped farm the dry corn and other staples. It has been a joy to watch Mike's continuing self-liberation from Society and the accompanying improvement in his sense for agriculture and for life. Mike is creating and heading up the Columbia River Watershed (CRW) satellite operation of Sunroot Gardens in 2010. He is seeking plots and partners for this NE PDX enterprise.

AND THE WINNER IS:

The Firepit Cat, whose photo is attached to this email. The Firepit Cat has spent countless hours hunting and killing rodents on the grounds and enjoying catnip. While we were all partying at OG Farm, at the Staple Crops potluck, she was back at greenhouse, protecting the stored grains from predation. Three cheers for the Firepit Cat !!! May she continue offering disemboweled corpses until our food is safe !! Yay, yay, yay, the Firepit Cat !

In the absence of a reaction from the Firepit Cat regarding the winning of this award (at least in English), we present an acceptance speech from Peter Salem:

"Hello sage members of the staple crops project. I regret that I could not be there today to accept the first ever 'Peter Salem Award for Excellence in a Seasonal Volunteer,' and wish everyone a delicious and plentiful meal. It's really an honor to have inspired this new competition, and would like to take this chance to bid potential future competitors 'bon chance'. I must say this is going to look very good on my resume, and it would look better if I won two years in a row. So don't count me out of the running for 2010, Farmer K, I'll be back this summer. But this is a time for celebrating the staple crops, an enormous project. I don't know the final status for all of the crops, but I had the opportunity to participate in the initial stages of the wheat harvest. As many of you know we started on a dry, hot weekend in the middle of August. On those days we scythed down the wheat, carried it to be threshed in a chipper-shredder, and winnowed in an apparatus made with a leaf blower. At the end of the day on Sunday at least two thirds of an acre still stood. I bet X and Angel I could scythe it down by the end of Monday, not really believing it was likely but really wanting to see it done. Of course there was still wheat at the end of the day, but I thought to myself as the hottest part of the day was passing that being there was one of the most honest thing's I'd done. There I'd stopped complaining about the social and economic structures that exist today and started working to create something different while providing for our most basic need. It was honest because my values and actions were less in contradiction. The staple crops project is an achievement, because it is putting social awareness into context. So I hope everyone brought his or her smiles, and I bid you all, 'bon appetit'. -Peter"

Thank you, Peter, for your gracious speech, and we look forward to your future attempts to win the award named for you. As Red Sox fans used to say, "There's always next year."

2010 Season: Love & Loathing

2010 brought another rotation of characters to the stage. Tom of Mall56 was out, moved to the East Coast where his girlfriend had been accepted for school. Angel was active for the first few months of the year but then suffered a shoulder injury that kept him off bicycles all summer. I had spent the winter in the Firepit greenhouse after a carpenter friend helped renovate the inside. It still leaked (argh!) but the addition of insulation meant a heater would actually function to keep it warm. On sunny days, the greenhouse heated up the building so much that I could hang out in a sleeveless t-shirt while the cold wind whipped through the bare branches of the fruit trees outside. Firepit Cat came and went of his/her own accord through a special door I made.

Demeter was still in town and had moved to a place called "Riverhouse" in Portland's Sellwood district, away to the south. Riverhouse hosted a CSA run by a woman named Clarabelle and 2010 would be her second season running that business. Clarabelle wanted to build a greenhouse on the property and because I wanted Demeter to make a good impression on her new friends, I had River deliver a bunch of windows from my stash as a donation. They were gratefully received. Demeter ended up closing down her CSA to collaborate with Clarabelle full-time.

Clarabelle was friends with another female urban farmer, Augusta, who ran an operation called the Westmoreland Garden Club. WGC was not a CSA—her customers were high-end restaurants—but she had a "distributed network" like Sunroot did. In 2010, she swapped out her business partner for a boyfriend named Lairy and added new plots. At her prodding, Riverhouse CSA, which had previously been located entirely at its home site, expanded its operation to other properties too.

Demeter's presence at Riverhouse brought me there for increasingly frequent visits as the weather warmed up and soon I had befriended

Clarabelle and Augusta and Lairy. Clarabelle had a degree in Environmental Science with a minor in Botany and Augusta had a degree in agriculture, so for the first time I was hanging out with farmers who had formal educations related to the work. I appreciated how their book knowledge complemented my self-taught field work, and that appreciation was mutual.

"Japanese Prison Josh," or "J.P." was another character who rose to prominence in 2010. J. P. had his own market-farm and CSA operation based out of a half-acre plot in outer-Southeast named "Alleisdair" (after a local cat, at the insistence of Angel and me). Inspired explicitly by the media stories about Sunroot Gardens, he had netted it in early 2009 and had cold-called me for help. For most of 2009, he struggled to translate his visions into reality and not much happened there until the late summer when Angel brought the tractor up from Milwaukie to till the entire property. The three of us spent a few highly productive days together planting up winter vegetables. In 2010 he was ready to get serious.

By late May, five of us—J.P., Augusta, Lairy, Clarabelle and I—were working together almost everyday on each other's plots and sharing tools, seeds, methods and ideas. (Demeter joined us occasionally but spent most of her time at Riverhouse playing the role of "farm-wife," cooking and organizing.) In terms of sheer "fun," the period of time that followed —May to July of 2010—was the peak of the Sunroot years for me; though we were all focused on the work with absolute seriousness, the atmosphere of our communal labors was light, even party-like. Sexual energy was always close to the surface—or riding giddily along on top of it—and to me that felt perfectly appropriate. After all, we were directly engaging ourselves in processes of fertility, pollination and reproduction. Hadn't peasants always enjoyed ribald behavior? Wasn't the springtime preternaturally imbued with raunchiness? Didn't a full life embrace the energy of lovers when lovers appeared?

But of course all good things must come to an end. By August, the scene had turned sour. Lairy claimed to be a recovered meth-addict, but still displayed the intense ego-mania of a speed-taker; more and more often, he could not settle down and play with others, but insisted on dominating. His patriarchal streak, at first subsumed by the power of the group, asserted itself and he became unnecessarily possessive of Augusta. She assented because she wanted to be paired-up. A toxic chemistry at first seeped but soon flooded out from the two of them. Accusations and insults became their currency, followed by subversion and theft. Our partnership ended acrimoniously. As is usual with such situations, the root cause—the socially-conditioned ego—was never acknowledged or examined, and ridiculous blame-game stories were thrust forward to bury the truth.

Clarabelle took the brunt of the abuse from Augusta. They had been best friends before—with Augusta serving as counsel to Clarabelle as she exited a marriage engagement with uncertainty and painful slowness— but now Augusta threw Clarabelle's words back in her face and tried to paint her as untrustworthy, unappreciative and unlovable. To me, it was plain that Augusta was describing her own perceptions of herself with these baseless accusations and that ignoring her was the only response that made sense, but Clarabelle did not see it that way at the time. She took it all on and suffered some anguish for a spell. Angel and I tried our best to pull her through, but she had her own lessons that she could only learn herself by plunging into some dark places first. Sometimes that's how life goes.

In the confusion of those times, J. P. claimed neutrality with his words, but took sides with Augusta and Lairy in his actions. Such unconscious duplicity is par for the course in our cruel and insensitive society, and he also had his own lessons to work through. In time, Augusta and Lairy moved on and we renewed our friendship with him.

For the remainder of the 2010 season, Clarabelle and I worked together closely and our CSAs were both well-provisioned for the most part. With three harvest days each week, and with pick-up spots located up to five miles apart, we found it easier to drive than to bike, and for the first time in Sunroot's history, two wheels took a back seat to four. I accepted this without judging myself. My knees, after five years of hauling loads, were constantly sore and I had concerns about permanent damage. (As of this writing, in October 2015, I can tell that these concerns were not misplaced. My knees have still not recovered from those days.)

I also accepted the change in transport because I knew that 2010 was the final season for Sunroot Gardens. I didn't have anything to prove anymore and just wanted to finish out the year. My interest in urban farming had waned as I saw that the revolution I had envisioned was not happening and didn't seem likely. Agriculturally, I wanted to work one piece of land, not 30 or 40 or even 5. Personally ("spiritually"), I desired a rural setting—with less noise, cleaner air and more starlight—to pursue my path.

04.05.2010: "Don't You Ever Take Time Off?"

Date: Mon, 05 Apr 2010 11:49:49 -0700
Subject: [sunrootgardens] Survey & News

DON'T YOU EVER TAKE TIME OFF? I don't need to take time off if I don't take time on! ;) A couple-three years ago, I took that coin that says "work" on one side and "play" on the other, and tossed it over my shoulder for good luck. Life With The Plants is outside the calendar, outside clock time. Watching the kale/collards overwinter and now burst into bloom, it is obvious that THEY live in the "real world" and WE (human society) live in a dream world. As important as we think our jobs are, our friends and family, our ideas and values and beliefs—they are all meaningless in the real world of the living planet. The idea that anyone HAS to do something (get a job, make a commitment, have respect, etc.) is pure unadulterated bullshit. Don't like what you're doing? Stop. Sick of your partner? Dump em. Don't feel like being superficial in order to be polite? Quit! There is no real penalty if you are being truly real, which is to be truly free...,

The rewards of joining the plants makes it no loss when people don't understand or when material luxuries are absent. The clarity of mind and ease of movement are their own rewards, beyond the value of money or the approval of ones' peers, both of which are ultimately (and immediately) empty. And the stomach gets to be full !!

— "People without plants are in a state of perpetual neurosis, a state of existential wanting." (Terence McKenna)

06.11.2010: "How To Eat Local & Enjoy It"

Date: Fri, 11 Jun 2010 10:11:05 -0700
Subject: [sunrootgardens] How To Eat Local & Enjoy It

Lots of folks in Foodie-Portland are concerned about how to eat well, and there's tons of approaches to the question, each of them outlining their own sets of goals and limitations. Some of them are quite dogmatic. All are overly complex. The answer to How To Eat Well is very simple:

Look around you.

What is vibrant and ready-to-eat in your local area right now?

That's what's healthy to eat.

Seasonal vegetables don't just taste good. They are offering to you a mix of minerals and vitamins and properties that your body needs at that time. The bitter greens of early Spring are perfect for detoxing the body after the lessened activity of Winter. The cooling qualities of cucumbers are perfect for summer's heat. The substantial sweetness of a Delicata squash is perfect for a body heading into colder days and longer nights. Conversely, the coolness of the melon is unhealthy in January, when the weather is damp, and the warming root of the parsnip does nothing for you in June, when the plant is going to seed and has become pithy.

I have been heard to say that I am not really a Farmer, but actually a Chef with extremely picky tastes, as in, Dammit-If-You-Want-Good-Ingredients-You-Have-To-Grow-Them-Yourself. At one time my set of kitchen toys was impressively broad, and my spice cabinet diverse. I had awesome cast iron and Calphalon, a set of bone-china, and dozens of wine glasses.

It is quite different, then, for me to be living with my current kitchen circumstance, which is comprised of: a one-burner induction hot-plate (on which cast iron won't work), one fry pan, one sauce pan, one chopping knife, one spatula, no plates or bowls to speak of, a handful of mismatched silverware, and an electric tea kettle. My spice rack has sea-salt, vinegar and oil—that's it! I pick fresh cooking herbs as needed. This is a bare bones, quasi-camping set-up. And guess what? I am eating the most enjoyable meals of my life because the ingredients are so damn good! When food is seasonal, fresh, and blocks-local, All You Have To Do Is Chop It Up and Heat It. That's it. Recipes are unnecessary. The things that are growing well at the same time happen to taste good together too. And to have complementary qualities, nutritionally.

And guess what else? Eating what's fresh, local and seasonal was The Only Option for People for, oh, the first two million years or so of human history. That is to say, there's nothing Special about it. The snobby Portland foodie-scene is pretentious-as-fuck, and seemingly unaware of the self-indulgent luxury of its "values" and "choices." It's only because we live in such a wealthy society, built on the rapaciousness of globalized capitalism, that something as ridiculous as Apples From New Zealand (like you'll find at the Co-op at this time of year) is even possible as a "choice." The resulting physical dis-ease, mental befuddlement, and lack of holistic self-awareness is obvious in the population. I know this one intimately from the inside out, from close observation.

Fortunately, there's an easy way out. Pay attention to what we're offering. Take a lot of it. Experiment. Don't worry about Taking Too Much. Bring compostable extras back to the Firepit if you want and I will bring them to grateful chicken friends. "There's nothing like chicken love," says Miss Clarabelle of Riverhouse CSA, and she's right. There's also nothing like discovering yourself—and your own bubbling spring of love—through your own senses, and Food is a way to do that multiple times daily. So-called "Transcendence" is readily available through your lips, down your throat, and deep down inside you. Pay attention to how it comes out the other end, too, as that also gives you clues.

COOKING IDEA / RECIPE:

We've got a Tsunami of Radishes about to hit, sometime in the next week. They will be a mix of delightful colored balls (and cocks), with the greens. The other day, a helper was over here and mentioned she was hungry. I made a pot of sticky rice, taking it off the heat after it was up to a boil and letting it finish under a wool blanket so I could have the burner for the next step. Quinoa or millet or any other grain would work here. Then I sautéed a bouquet of sliced radish roots in oil, with sea salt. When they were getting cooked through but were still crisp at heart, I tossed on the chopped radish greens, threw in a splash of red wine vinegar, and put on the lid. I turned off the heat after a minute, with a towel over the top to hold in the heat. In another two minutes, I pulled it off, set it down in front of sister, with the rice, and wished her Bon Apetit. She dug in, and was Astounded at how delicious it was. How sweet the radishes, with just a hint of their spicy heat. How tender and bright the greens, with the slightest "al dente" in the stems. How clean and refreshing the simple spicing. She gobbled up a bunch with a silver fork, saying not a word, because why distract the eating mouth with words? Eating is no time for multi-tasking. Eating is a time for eating.

Bon Apetit !!

06.14.2010: "Download re. Community"

Starting in June of 2010, produce pick-up days were happening twice a week, Tuesdays and Fridays. I dubbed Fridays as "Farmy Fridays" and tried to create a scene that night. I had always enjoyed throwing parties, and this weekly *fête* was intended as a meet-up for anyone and everyone interested in urban farming so they could find ways of cooperating in the present and future. Since I was planning to be gone in 2011, I was hoping that various people would pick up the reigns. I was well aware that my own success was in my network, so I wanted my network to network within itself and to be fruitful after I left. Here was the announcement of this concept:

Date: Mon, 14 Jun 2010 12:36:59 -0700
Subject: [sunrootgardens] Produce 2x weekly

Starting on Tuesday, July 15th, Sunroot Gardens will be offering produce TWICE weekly.

Where: Firepit Garden.

When: 4pm 'til nightfall (11ish pm these days)

Who: $-subscribers, landlenders, helpers, Friends-of-the-Farm, etc.

Fridays are "Farmy Fridays." This scene is all about as many folks as possible meeting each other. Who knows for sure what the future will bring, but a dollar sez that we will be breaking up the Sunroot Empire over the course of this year and handing out the pieces to whomever wants to take them over for 2011. On a "Farmy Friday," you could meet land-lenders, $-subscribers, Folks-with-Skills (construction, welding, engine-repair, etc.), helpers, Future Urban Farmers, fun summer hook-ups, and more. Expect food, drink, fire, plotting, and monkey business. With this ongoing event, I am intending to offer fertile ground for seeding Tomorrow's food-based-communities in this City.

Rec'd Download re. Community the other day, and here it is: Community is the Active Cooperative Response to Need. That's it. Attempts to create community will fail when focused on anything not Needed, such as a religious or political philosophy, a common but spurious interest, or a recreational activity. I would go so far as to say that community is *not* created or organized at all, but is simply Found.

So what is "Needed"? Food, Water, Clothing, Shelter, Child-rearing.

Activity for supporting these needs can be focused purely on logistics, and their utilitarian/practical aspects. "Mission Statements," "Guiding Principles," and "Consistency" are merely intellectual constructs and are unnecessary for accomplishing work. These mind-based superficialities lead quickly to puritanism and rules. Rules are an attempt to give a "Yes" or a "No" to a situation ahead of time. This gets so tricky so fast because we can never foresee ahead of time what all the factors will be that we need to consider. Then, with the Rule, we fail to see clearly the situation for what it really is—because we are trying to make it fit the Rule—and we end up making choices based on principle instead of the facts. The results of this regimented, rule-based Society are clear all around us in the pharmie-drugged population, which is nearly universally unhappy, sluggish, and irresponsible.

So Farmy Fridays are for folks to find folks to cooperate with around the cultivation, harvesting, processing and ENJOYMENT of Food here in the City of Roses, Cascadia. I am thrilled to introduce people, share my knowledge and experience, and cook. C'mon out and see what you find. Dollar sez it will be things you have only dreamed about.

I must add a correction to this. I wrote: "Attempts to create community will fail when focused on anything not Needed, such as a religious or political philosophy, a common but spurious interest, or a recreational activity." This is not factual. Agricultural communities based around religion have been famously successful: think the Amish and Mennonites. Indeed, I believe the future of the USA will prominently feature theocracy as a form of governance and community, certainly on the local level and perhaps (frighteningly) on the national one.

Churches are well-positioned to react to logistical changes such as food shortages, resource interruption and disasters. Their members meet once a week, recognize leaders, and are willing to submit their individuality to the collective. Churches as legal entities are also property-owners, many of them with plenty of space to raise vegetables. I know of more than one Seventh Day Adventist parish in Oregon that holds multiple acres, which will enable them to get serious about staple crop production, too, if they choose. (I once put on a seed-saving workshop for a Seventh Day Adventist Church in Newburg and found the folks not only interested but attentive and focused.)

The lefty atheists and agnostics will (and already do) have challenges cooperating with each other. They distract themselves with ego issues under the guise of identity politics. They put their intellect before their intuition to the point of denying the existence of the latter. They view the world as meaningless (because the meaning*ful* character of the religious believer's view is so preposterous) without understanding that their knee-jerk reaction is merely a cynic's foil, and that they are denying themselves an awareness of the inherent, autonomous vitality of existence.

"Community," in its true form—neither based on religious obedience nor on secular conceptuality—is probably beyond the ability of the current generations in the USA to discover. Natural disasters can bring out something that resembles it, but these are temporary states of affairs. The real test will come after we collectively enter what James Kunstler has dubbed, "The Long Emergency"—a period of permanent economic contraction, resource depletion and social dissolution—*and* when anyone who denies this reality is roundly derided by their peers. This is not a period whose commencement I wish for; much suffering will mark those times and not only for humans. "True" community might not coalesce until the Long Emergency has been underway for decades. Honestly, I do not expect to witness it in my lifetime.

In-Depth – "Skip work and come play": All About Volunteers

"At this time of year, we are busting out beds everywhere. If that sounds like fun to you, then come join us!" (Sunroot Gardens email, April 22, 2010)

Sunroot Gardens, like the other urban farming operations in those days, relied on volunteers to help with all aspects of planting, tending and harvesting. By "rely," I mean that our visions and plans counted on someone besides ourselves to show up and put in some labor.

During my activist days, I had dealt with all the issues of trying to recruit, train and work with volunteers. My skills in these areas had improved with practice—which included being a volunteer coordinator for the Ralph Nader presidential campaign in 2000 in four cities—and when I began urban-farming in 2005, I daresay my abilities were above average. Certainly, once willing volunteers were put in front of me, I could effectively size them up and suss out where to place them in the workflow. Fundamentally, I enjoyed teaching (and still do).

A large percentage of the messages sent out to the Sunroot email list were calls-for-helpers. I didn't have a stock spiel and I varied my approach as I felt appropriate. I was always trying to attract people's attention or address their tastes in different ways.

This message, sent April, 2008, gives a day-by-day schedule of volunteer opportunities. I include it here as a sample of its type.

Date: Mon, 21 Apr 2008 09:10:59 -0700 (PDT)
Subject: [pdx-urban-csa] April Full Moon Farm Bulletin

Folks have been coming out to help on different projects, and everyone has been enjoying themselves. A real cooperative spirit has been in play. Check out the schedule below, to see where we'll be this week.

Monday, April 21

* morning: Bed prep and onion planting at the High House, 50th & Belmont

9:00 meet-up at the Firepit

* afternoon: Bed prep and carrot seeding at Mateo's, 48th & Caruthers

1:00 pm meet-up at the Firepit

Tuesday, April 22

* A day at Terwilliger Community Farm in SW

Weeding and thinning peas, other bed prep

9:30 am meet-up at the Ugly Mug Coffee Shop in Sellwood, on 13th just north of Tacoma

Wednesday, April 23

* Taters at Standing Stone, SE 122nd & Flavel

Yeah, that's "out there," but it's quite a spot. On the NE side of Mt. Scott, with a view of Mt. St. Helens, and a giant "glacial erratic" boulder standing at the foot of the slope

Call re. meet-up time to coordinate transportation, which might involve a vehicle to move all the seed potatoes out there.

Thursday, April 24

* A day in the Clonglutch, a set of five gardens along SE Long between 29th & 31st Aves: beet & turnip planting, and bed prep.

* 9:00 am meet-up at The Funky Door Coffee Shop, at SE 28th & Holgate

Friday, April 25

* Firepit work-party, all day.

Bed prep, weeding, chip spreading

Come anytime, but feel free to call first.

Work days at the Firepit often end with dinner cooked up by Mrs. K.

Saturday, April 26

* Lucky Cats! work-party, all day.

Bed prep, seeding, strawberry-planting, bird-watching

Meet at the Ugly Mug at 9 am

On any of these days, if you want to come along after the meeting time, you are still welcome Just give me or Melanie a call to see where we are and how to get there, etc. Bring a little food for yourself to eat there if you'd like. We send folks home with produce that's in-season and available.

As you can see, we were always moving. "Kollibri" is Norwegian for "hummingbird" and isn't this how those highly active, quick little birds behave?

People often offered to come help "sometime." They'd say to call them up "anytime." They wanted a personal invitation. Angel and I would tell them we were working "all the time," so it was up to them to show up when they wanted to. What we could do was keep them up-to-date through the email list. This note spelled out what to expect.

Date: Wed, 21 May 2008 09:08:31 -0700 (PDT)
Subject: [pdx-urban-csa] CSA shares ! / A note to helpers

A NOTE TO HELPERS:

We are doing farmwork all day every day through October, at locations all over SE. There are currently 99 people on this list, and that number is increasing. Personal invitations, therefore, are rarely made; this list is the sole notice of opportunities that you are likely to receive. Feel free to call Farmer K or Sister Melanie anytime to find out more details about where/when we are on any given day. If you have a particular day/time you would like to come out, just let us know, and we will tell you where/when we are.

Any amount of time helping is helpful. A half an hour of weeding is fine. So is a whole day doing the circuit. Whatever you would like to do will be welcomed. Meals are generally provided through the pooling of resources; you need not show up with your own food. Those people around at the beginning of the day are often offered breakfast; those at the end, dinner. Fresh produce will be sent home with you when available.

Previous skills and experience are unnecessary. Together we will find something to do that suits you. People often come away from the experience expressing satisfaction and a desire to return.

Any amount of participation is appreciated. This is your invitation.

This was the soft-pitch approach, easy and open. Fine if you're a self-starter, but not many people are, as we came to find out.

What follows is another kind of invitation. It takes the form of a journal-like entry describing current life-on-the-farm and was intended to woo people by illustrating the rewards accompanying the labor.

Date: Thu, 29 May 2008 21:44:48 -0700 (PDT)
Subject: [pdx-urban-csa] Sunroot CSA skipping a week / Farm News

> FARM NEWS: For the last 10 days, we have had five people working full time (50+ hours / week) plus other helpers. "The Partners" these days are me, Mel and Angel. Troy and Peter have been working full time just about every day in exchange for room and board, which is an ancient agricultural arrangement. They get breakfast and dinner cooked for them by me, lunch provided by donated food, plenty of produce, and a dry place to sleep—in the greenhouse on big burlap coffee bags from the Stumptown roastery, same as me. Daylight hours = working hours.

> This morning, we had fresh farm eggs scrambled with Shitake mushrooms and Walla Walla onions topped with goat cheese and served on olive bread. I grew the onions, I traded some of the onions for the eggs and mushrooms, and the cheese and bread were donated. In a restaurant, this would have been a $10+ item, although you can not find a restaurant in town with ingredients that fresh and high quality. "Living Large," we deemed it. A good life.

Good food for hard work in an egalitarian setting; not everyone's cup of tea, of course.

This next pitch placed farmwork within global and historical contexts. I was trying to illustrate how Sunroot fit into the bigger picture of the day's issues and how we could move forward by reaching back. Some people are inspired by the idea of participating in something that is larger than themselves. I hoped this message would entice those folks.

Date: Tue, 10 Jun 2008 19:07:57 -0700 (PDT)
Subject: [pdx-urban-csa] Crop selection changes due to World Food Crisis

This is a very busy time of the year! Over the next two weeks or so, anything that's a bean, a squash, or corn has to go into the ground, in order to have time to mature. So this includes not only produce for the CSAs, but also for the Staple Crops Project.

The Staple Crops Project, for those who don't know, is an attempt to raise food you can actually live on, being that fruits and vegetables aren't enough (for most people). Crops planted this year include: soup peas, dry beans, and quinoa, with soy beans and oilseed sunflowers and more coming.

The World Food Crisis is making the need for the Staple Crops Project more obvious. In the light of the global situation, which seems likely to strike what is called the U.S. in short order, I have made some changes to our crop planning for the CSAs.

Watermelons and Sweet Corn have been dropped. Both are delicious and have their nutritional qualities. Both also have a poor ratio of productivity to space, meaning they don't put out much for the room they take up. They are being replaced with soup beans, overwintering greens, and soy. Additionally, under-performing crops already in the ground—such as the leeks at Lucky Cats!—might be dug under and replaced with something heftier.

HELPER CALL-OUT We are not able to put out a weekly garden schedule right now because there is so much to do; decisions about where to go and what to do are being made on a day-by-day basis. However, we really do need as many helpers as possible, all day, every day of the week. If you have any time at all over the next two weeks, please give us a call and we will happily work you in. We will likely have more than one crew working simultaneously in different locations.

THIS IS CRUNCH TIME FOR PLANTING! What we get in now—esp. the Staple Crops—is what we will be living off of later.

Back in the old days, communities would come together around important parts of the agricultural year, including plantings and harvests. It was understood that everyone needed to pitch in for the common good at such times. The lifestyle has not been like that in the U.S. lately, but is likely to go back that way. We're starting that now, this year, here in this community.

This is when the rubber hits the road!

(The bike tire rubber, that is.)

Thanks much, and see you soon.

"Ah, the good old days," right? I am reminded of the Billy Joel lyric in his song, "Keeping the Faith": "The good old days weren't always good / and tomorrow ain't as bad as it seems." So true.

Farmwork can be a year-round endeavor in Oregon due to the mild climate and this email made a call for helpers in early December. It also expands the scope of what Sunroot Gardens was offering to people interested in urban farming, specifically, the opportunity to start up their own operations if they were so motivated. We wanted to facilitate the expansion of urban farming as a movement in Portland. We had no desires to keep everything to ourselves. We hoped to help spark a vibrant scene with not just dozens, but hundreds, of urban farmers who could transform Portland into a veritable "garden city" where you couldn't bike down any block without seeing vegetables in front yards. Only then would "urban farming" be a real thing and not just a media-manufactured meme.

But if our offer was to empower people become their own bosses, we also made it clear that we were referring not merely to starting one's own business; this was also an invitation to step outside the box of conventional society and discover other ways of experiencing life: to foster one's own personal ("spiritual") growth.

It was with these things in mind that I sent this email, though I also stressed that our offers, being serious, could only be accepted by people who were serious.

Date: Wed, 3 Dec 2008 11:59:29 -0800 (PST)
Subject: [sunrootgardens] Winter work continues / New CSA year begins

Hello all, Farmer K here:

*** HELP WITH WINTER STUFF *** We are taking advantage of non-rainy windows in the weather to continue putting-beds-to-bed and getting new gardens put under cover-crop for the winter. This is outdoor work, in some mud, with hand tools and with mechanization (tillers, a tractor), doing work now that will make spring/summer that much easier and more productive.

We hope to do major prep work Thurs, Fri, Sat, Sunday.

*** Becoming an Urban Farmer Yerself *** If you are interested in becoming an urban farmer yourself, this is an opportunity for you to learn more about city gardening, including that most pleasing of activities: Taking Out A Lawn, by many different means. If you are interested in sharing plots for next year for your own urban farming operation, this is your opportunity to see what's available. I am entirely open to splitting various plots for the next year, or even giving the majority of space of a few of them, over to the care of other people if they are serious. This is the big phrase: "if they are serious." Merely vocalizing the desire, "I want to do urban farming" means little-to-nothing to me, and does not represent seriousness by itself. Serious farming does not happen as the "second fiddle" to other activities, but is central and primary. "Vocation" is a good word, and I thank David Ashton of the Sellwood Bee for that one.

The priorities and cycles of Farming are not merely different from the priorities and cycles of City Life—they are At-Odds with them. As I have immersed myself in these things, I have felt the City recede further and further from my consciousness. Without physically leaving the City, I have found myself further outside of it. I now witness it (the City and its cultures and customs) as if through a telescope turned backwards; in focus but far away. Interestingly, this has gone a long way towards gratifying previous desires to get-out-of-the-city-and-go-to-the-country, a sentiment I hear often from other people.

Though to give credit where credit is due, my reconciliation with City Life can also be put at the padded feet of this town's cats, who—through their example—have taught me much about being "chill" in the city, unperturbed by its tumult.

All this being said, people who want to learn about farming in the city within the Sunroot Gardens context are free to spend as much time doing so as they would like. I offer no promises whatsoever as a "Teacher." The onus lies with the "Student" to learn or not learn from what they experience. And in that process, the roles can disappear. (Perhaps in such disappearance is the only real learning/teaching.)

Call or email.

Not all calls-for-helpers were so grand in scope. In the spring of 2009, I offered this simple opportunity: to help collect compost from local businesses. It was soon accepted by a woman in the neighborhood who genuinely enjoyed it, often accompanied by her young children on their own bicycles. Since that time, the city of Portland has stepped up its composting services, so neighborhood-based initiatives like this one no longer have the usefulness they did. It's true that more tonnage is now being composted, but it all goes to a central location outside town and is sold back to city people in plastic bags. Something was gained and something was lost.

Date: Tue, 3 Mar 2009 11:33:55 -0800 (PST)
Subject: [sunrootgardens] Sowing oats, getting busy, step up!

Grey morning greetings from Farmer K:

The weather for the next 7-10 days is predicted to have highs in the 50s and regular precipitation.   This is perfect planting weather! We plan to seed like crazy all over town over this period. Every day of the week for the next 7-10 days we will be somewhere, doing something. We are not putting out a schedule of when/where because it would be a waste of time and energy to try to make such predictions, and then to live up to them.

To get involved on any particular day, contact Farmer K at 503.686.5557, or drop an email at least the day before.

...2) Join the COMPOST BRIGADE!! Farmer K has been setting up local cafes with buckets for collecting their coffee grounds and kitchen waste. So far, we have Fine Grind (39th & Lincoln), Common Grounds (Hawthorne @34th), and Red Square (45th & Belmont). At current rates, we need people to stop by these places 2-3 times a week. I have set up a special bike cart at the Firepit Garden (which is where the compost should come to) for hauling empties to these places, and bringing fulls/half-fulls back. This is free for people to use if they are interested in helping out this way. Swinging around to the three of them and bringing it the Firepit is a task of less than an hour. Someone could also do this with their vehicle if they would like. Getting the compost is most important—not the means of transporting it.

For becoming a Compost Brigadier, you will be able to get fresh produce. Until May or so, this will happen on a case-by-case, on-the-spot basis. From June or so onwards, being a regular Compost Brigadier will earn you an invitation to the Community produce distro day, which will happen at the Firepit Garden once a week, likely on Fridays. This day will be set up just like the day for the $-paying CSA members: The current produce will be laid out in a farmers-market-style set up and folks will be free to stop by between x & y time to get what they want.

We will be continuing to set up compost pick-ups with cafes/stores all over SE this spring, finding labor/tasks that need people to do them. That is my methodology here. I am by no means limiting the scope of Sunroot Gardens to what I myself personally can do. Instead, I am finding the labor/tasks that need to be one, and putting the word out. This is how our "community" will form—through the labor that needs to be done around growing food.

Here, farming as path-to-personal-("spiritual")-liberation is again presented, but also an important reminder about the role of children in Sunroot agricultural work. We had to stress, over and over again, that coming out to help with farming didn't require finding childcare: children were assets, not impediments, to the new communities we hoped to see flourish.

Date: Thu, 26 Mar 2009 23:01:22 -0700 (PDT)
Subject: [sunrootgardens] Fourth Moon Farming Update -
Helper opps

...Urban farming provides escape from the city-as-social-structure without leaving the city-as-built-environment. The plants don't give a flying fuck who the president is, how Wall Street is doing, or whether that cute person from the party is going to call you back. As such, none of those things have anything to do with what you're doing while you're farming. 'tis an opportunity to live more freely.

...BTW - CHILDREN ARE WELCOME to any garden anytime. Gardens are good for children and children are good for gardens.

* Many farmers got their start young.
* Eating dirt leads to higher immunity.
* So does eating the fresh veggies.
* Kids enjoy being around adults who are enjoying themselves.
* For goddsake, don't they spend too much time inside?!?!
* No Helmet Required.

With Sunroot Gardens, I was trying to develop new ways of doing business that weren't focused exclusively on money. I believed (and still do) that we must invest ourselves in co-operative efforts motivated by a sense of the common good in order to change the destructive course of humanity. In that spirit, starting in the summer of 2009, I started offering produce for pick-up on a second day each week devoted entirely to helpers and people who wanted to barter. The customers who had invested money for CSA shares would still have their own pick-up day and this add-on was intended to highlight the importance of non-monetary contributors and to reward them equally. I have never heard of another CSA doing this.

Date: Wed, 10 Jun 2009 08:52:24 -0700
Subject: [portland-urban-farmers-coop] Free Produce Friday 6/16, @Firepit 4-Sundown

Waxing moon greeting from Farmer K here with an invitation:

This Friday the 12th, from 4-sundown, is the day for Helpers, Land Lenders, Bike Mechanix, etc., to come get produce from Sunroot Gardens!

HOW THIS IS SPECIAL

As far as I know, Sunroot Gardens is the first CSA to add a distribution day for non-monetary investors. Every other CSA I know gives produce only to the $ investors. Here at Sunroot, we recognize that labor, land, etc., are AS important (if not more so) than $, and that the farm cannot farm without them.

...With the value of $ in fluctuation, and promising to become more unstable, BARTER will become more prominent in our daily lives. This is merely a typical feature of an unraveling empire, and is nothing unusual. By relying less on $, Sunroot can make itself a more sustainable business in this context. That's an actual "sustainability" not a jargonistic one like in the corporate adverts.

This is a hang-out event. Bring a bottle (or five). We've got home grown tobacco for those who want to try it. We'll also be finding homes for tobacco plants for this year's crop if you're interested.

If people get hungry, Mrs. K can make some dinner. Meet the other lovely folks working with Sunroot.

Come to the Firepit on Friday for the Farmy Scene!!

By mid-summer of 2009, it was clear that the boost given to Sunroot by the media was wearing off, at least in terms of numbers of helpers. We also understood that 2008 had been marked with a particular exuberance in Portland due to the end of Bush's presidential term and the ascendancy of Obama, with his "hope" and "change." In 2009, these collective delusions were wearing off and idealism was on the down-turn again. Sunroot was no longer perceived as being fresh and exciting. And if the media were to be believed, "urban farming" was a full-on, burgeoning enterprise in Portland. With that picture taking hold in people's minds (false as it was), the sense of urgency also passed; "that issue" was being handled and everyone could go back to what they had been doing before. I saw how the media coverage, though positive on its face, was a double-edged sword. Not that this shocked me; my years of Indymedia activism had educated me about the corporate media's nature.

The following email addresses the declining number of helpers. Other volunteer-dependent organizations would probably never dare to speak as candidly as I did here for fear of alienating people or painting their organization in a weak light, but for me the entire Sunroot operation was an experiment in honest, open living, so I did not hesitate to put my thoughts forward. Weren't these the issues we needed to work through together? If so, there was no point in denying them.

Date: Tue, 16 Jun 2009 11:05:38 -0700
Subject: [portland-urban-farmers-coop] CSA share this
Wednesday / Work Party on Saturday

NOTE ON FARM HELPING

there have been far fewer farm helpers this year than last, even though the need to grow food at this time is arguably more important (what with the conventional $ economy collapsing and the corporate food system crumbling). What this means is that there is more food available for those people who do come out to help. I say this just so folks know that the labor-for-food offer that Sunroot makes is serious. Besides the every-Friday-distro day that will be starting up soon, people can also get produce on the spot when they help.

I feel like there are two reasons for fewer helpers this year:

1) We have not been scheduling many work parties. The complexity of this distributed farming project, and the changeable weather

have made that impossible. Therefore, we have been depending on those people who are SELF MOTIVATING to volunteer themselves when they want to help. Apparently, not many people are self-motivating.

2) The "malaise"—for want of a better word—of the culture at large. It seems like people are getting hopeless and sluggish as the status-quo changes. Used to luxurious living after 6-7 decades of oil-produced wealth, people don't know what to do now that they will have to start taking care of themselves again, like the human animal did for itself for several hundred thousand years. So they are doing nothing, or doing something at odds with their best interest as a spirit living in a body.

90+% of the society is doing nothing in DIRECT and MATERIAL support of itself (ie. growing food), but rather has been using INDIRECT and ABSTRACT ways for a couple generations (ie. money, insurance, etc.).

Now, as the party is ending, I see a self-destructive energy running through the culture. There seems to be this assumption that everything will somehow be okay, and that people will be able to feed/house/clothe themselves without making any changes. This concept is ludicrous, of course, but the subject is generally forbidden. People seem frozen in place, unable to act.

That's their (your?) problem.

And honestly, it's none of my business.

I don't personally—me, myself—care if people come to help with the Sunroot farming project and learn more about how to take care of themselves. I am figuring it out for myself, and that's all I can do. The offer is there for people who want to come out, but if I spent every day alone for the rest of the season, I would also be fine with that. Would less get grown/harvested? Absolutely. Would I be able to grow enough for me? I have no doubt.

I have no complaints about any of this. When I look at the world, I see it as absolutely perfect, just as it is. I am enjoying myself more deeply and purely these days, personally, than I even thought possibly in previous times. Urban farming at the end of the American Empire—what a gig! This process of self-liberation from the trap of Culture has given me joy unspeakable. I feel like the plants have been my gateway, or that they have shown me that the gateway has always been within myself, and that there has never been anything to look for anywhere else (or with anyone else).

I have no idea if helping with urban farming will lead to a liberation like this for any of you, but you are certainly always welcome to join us in the field and find out. You have nothing to lose. You never do!!!!

From my years of experience working with activist groups and volunteers, I knew that prospective helpers could be hesitant to make obligations and also had fears about making mistakes. In this email, I tried to relieve people of both concerns. I also made an offer that, looking back, I find astounding. At the time—Summer 2009—we were trying to establish a new farmers' market exclusively for urban farmers and needed help staffing the booth. Read on:

Date: Fri, 10 Jul 2009 18:19:15 -0700
Subject: [portland-urban-farmers-coop] Sunroot Gardens
FARM NEWS

...OBLIGATIONS FROM HELPERS There are NONE. Show up and you will be offered produce. There are no mistakes. Broke a tool? Stepped on something? Picked something at the wrong time? Whatever. It happens. Anyway, it just did if it did. While there is LABOR involved with farming, with Sunroot it is not WORK. The work/play dichotomy is a coin best tossed over the shoulder for luck before moving on. There is no "making a living" for me in this, there is just "living," and this is what I have to share: An opportunity to be around vibrant living things and eat them. As "something to do," farming does not compete with other "entertainment options"; hence I do not bother to try to SELL it as an activity that, a) will be "fun," b) will salve your desire to "save the world" (such nonsense is none-of-my-business), or, c) is necessary to the running of Sunroot (which will be its own thing regardless of whether any one particular person shows up or not).

You are essential to no one and to no thing besides yourself. Sunroot being outside with other self-liberating folks around delicious food; an offer to drop all social obligations and just enjoy yourself, essence-ially. Come and Go as you please. With no obligation. 503.686.5557

FARMERS' MARKET HELP Sunroot has been offering produce from a booth at the brand-new Hawthorne Farmers' Market. Sundays 1-6pm. Tom of Mall56 has been covering it so far, but will be out-of-town or of limited availability for the Next Three Markets. So, here's what we are offering: If you come to the market and handle the distro of produce from the Sunroot table, then you get to keep anything you receive. This market is cash/barter only (no plastic card machines). Tom has offered some $ to the farm after the markets, from the day's take, but has been under no obligation to do so.

We will bring tables, tents, produce, etc., to the Market (by noon at the latest), and will help set up and take down. Just looking for folks to staff the spot. The market days so far have featured other urban farmers, music, hot bikes and cool kidz, among other things. A nice little scene seems to be getting going. Interested? Call Farmer K.

So, in exchange for just showing up, market helpers would receive *all* of the proceeds from that day's sales? Not a percentage, not just a box of produce, but every dollar and cent made, *plus* a box of produce? Wow. This is how serious we were about wanting to create a re-localized food economy based on cooperative effort. As I said above, in retrospect I find this offer astounding.

Believe it or not, I seem to recall that only one person ever took us up on this offer. Was it any mystery that my disillusionment with helpers and the social climate grew apace? Or that I could no longer deny that our visions of a revolutionary urban farming movement were just that: visions, the stuff of dreams, with no substance in the material world. For these and other reasons, I knew by the autumn of 2009 that I only had one more year of urban-farming left in me.

It should come as no surprise, then, that the following call-for-helpers, issued in early 2010, is far more staid in tone than those that had preceded it. We were still making offers far more generous than other farming operations, but I was also upping the sense of seriousness. There is also much here that attempts to disarm the potential helper of their own fears. In retrospect, I can see that I was probably going over most people's heads.

...Sunroot Gardens does not count on helpers, but that is only because we have learned not to count on anything. We invite you to relieve yourself of your own expectations when you come out to the field. (In a compost pile is always a good spot.)

If you want to commit to working two days a week, or one morning, or everyday for a month, that is your business. We will not hold you to it, so you can skip any apologies if you don't follow your self-imposed schedule. The same is true of being "on time"; we don't wait on people—we just move as we move, and you are free to jump on the train wherever and whenever you'd like. If you're keeping track of hours for a school project or something, we will sign what you put in front of us. We can also write a letter of that kind if requested.

If we see you being careless with tools or lifeforms that are important to the farm, we will stop you.

When calling for a location, please skip the "how are you" formalities. Just say why you're calling. I have experienced physical pain from cell-phone use and work to keep conversations on them short. If it helps, pretend we're on walkie-talkies, using the brief, snappy phrases characteristic of that medium.

If you are around when Mrs. K is preparing food, you will be offered some, merely for being present. Indeed, it's the food that got me interested in farming in the first place. We often bring food and water to work sites to share; feel free to bring your own. These spontaneous combinations of what-everybody-brought and what-we-just-picked have resulted in some fine field meals!

Biking is often the easiest way to keep up with us. The current garden network was discovered and developed by bicycle, so it has a bike-logic to it. We have also been grateful for the automobiles that have been made available to the farm for moving heavy or awkward things around. We try to keep an extra bike or two on hand at the Firepit Garden, and sometimes people have bussed there and then gotten on a bike.

If we see a cat who might appreciate catnip or an ear-scratch, we will feel free to stop and investigate, even if we are "in a hurry" or "running late."

I am happy to teach what I know to people who want to learn. I don't have any trade secrets. If you feel like paying attention and applying yourself, Sunroot Gardens offers more than any farm internship program that I have heard of, because the level of your involvement is self-directed. In the past, we have had helpers who have chosen to specialize in a particular area to such a depth that they have ended up teaching *us* what they learned. Some helpers have invested so much time that they gain "farmer status" themselves, which entitles them to a "farmer's share" of a particular crop or crops. These are folks who are contributing more than just time, bringing their own resources to a project, be that seed, money, specialized equipment, etc. They are considering the crop "their own," too. The more you contribute, the more you get back. Last year, there were 5 farmers for the height of the season. Currently, we are at two, plus one up north at the Columbia River Watershed (CRW) satellite operation.

We welcome any level of commitment, even if that is just coming out once.

I have been seeing that farming is only somewhat about What-You-Know. Mostly, it is about What-You-Notice. As a result, I find that much time with helpers is spent simply trying to encourage people to PAY ATTENTION to what they are trying to do. This of course flies in the face of American culture, which is focused on sense-deadening, narcissism, and obedience. On occasion, I have demanded silence during farmwork, in order for people to find focus on the labor. I have also made people leave when they are insistently disruptive.

The cycles and priorities of Farming bear nearly no relation to those of Modern Society. As such, getting into Farming offers the opportunity to free oneself of the strictures of that society. With the observation that the kale doesn't give a shit about who's president, can come the realization that it doesn't matter for you, either. Politics (and religion and culture) are their own melodramas, with themes and conflicts that have not changed over the millennia. A quick perusal of Shakespeare, Greek drama, and the history of rise/collapse confirms this. Engaging oneself in Farming is a way of being part of the *real* Real World. You can't eat an insurance policy.

The joy I have regularly experienced since turning to full-time farming is unlike anything I ever felt in the conventional work world. It has allowed me to take the coin that is labeled "work" on one side and "play" on the other, and toss it over my shoulder for good luck. Farming is not "work" for me; it involves *labor*, but that is different. "Work" was following someone else's rules, trying to pretend they were important to me, and the measuring of "success" by arbitrary superficialities. The cruelest con is the self-con, and modern society—with its media, political institutions, and ridiculous family/friend roles—is focused on the constant self-con. The plants in the field don't give a fuck about any of that, and the degree to which we focus on the virtual rather than the actual ends up handicapping us in efforts at community-self-sufficiency such as farming.

I put all of this out there in order to weed out those who are not really interested in helping. After a few years of this activity, I no longer take it seriously when someone says they want to help with farmwork. When I hear "I want to help," I often interpret it as "it is important to my sense of self-image to present myself as someone who wants to help." Again, the veggies don't care about that.

So, if you are still interested in helping, give me or Angel a call, and we will tell you what is going on.

The "Columbia River Watershed (CRW) satellite operation," mentioned above, was a spin-off urban farming enterprise started by one of Sunroot's most active helpers in 2009. We provided him with as much as we could: land-leads, seed, amendments, starts, tools, construction materials and more, all free-of-charge. Sunroot had amassed so much wealth in these areas that we could spare some to a person with his seriousness. This act of generosity was never audited, but we received regular updates, tasty produce and saved seeds.

The idea that there were no-strings-attached was probably more important to me than it was to him. I was trying to actively live the ideal of "creating the world you want to see." I found that this effort had its own rewards for me, regardless of the experiment's results. The CRW satellite operation was one of the rare instances when someone took us up on our offers and made a real go of it.

In 2010, people were still asking me, "When are you farming?" as if I had an 8-5/M-F schedule. This email is a response that brings us full-circle back to the soft-pitch, though with a mischievous air.

Date: Sun, 09 May 2010 08:47:36 -0700
Subject: [sunrootgardens] Note for wanna-be-helpers

Long Day Greetings from Farmer K:

Check it out: it's about six weeks until the Summer Solstice, so the following three months are the longest days of the year. By late June, sunrises will be before 5:30 am and the sunset after 9 pm.

These long days will be filled with farmwork. Every day of the week. All day long.

I mention this because people often ask what days and what times we are working on farming activities. The short answer: all of the time.

So, if you are interested in helping out sometime, in order to learn more about farming, or to score some produce, or to meet hot farmie types, it is only your own schedule that you need to work around, not ours.

We bike around, play in the mud, eat fresh produce, check out bugs, plant seeds, water and weed, and dig. Dig?

You will have a better time with us than with your regular friends. You will wonder how you were ever satisfied with a conventional life. Skip work and come play with us.

Taking a totally different tack, I announced an urban farming course that would cost money for students. I had often received the suggestion. The concept was that some people would rather pay than volunteer, so not only would we get helpers, we charge them for the pleasure. Previously, I had resisted the idea; I believed that knowledge shouldn't be capitalized. But by the Summer of 2010, my feeling was that urban farming was so important that I shouldn't reject any method of spreading the word. So I tried it out. It's quite a syllabus that I offered:

Date: Tue, 18 May 2010 08:21:03 -0700
Subject: [sunrootgardens] Sunroot Gardens offering Urban Farming course

Bowing to popular demand, Sunroot Gardens is offering an Urban Farming course this season. Details follow....

DESIGN IN ACTION

A Ten Session Course in Urban Farming and Small-scale Grain Production

When: Every other Saturday, June 5 to October 23, 8am-3pm

Where: Various farm plots around Southeast Portland, plus suburban & rural country acreages.

Topics to be covered:

* Year-round vegetable growing and harvesting in the Cascadian bioregion

* Staple crops: Planting, cultivation, harvesting and processing. Crops to include oats, triticale, buckwheat, millet, quinoa, dry beans, flour corn, flax and more

* Medicinal herbs: cultivation and processing

* Farming-by-bicycle: challenges and rewards

* Starts and transplanting

* Design strategies and choices for plots of varying sizes, from small front yards to acreages

* Seed saving and basic plant-breeding principles

* Insect identification and management, including basics of bee-keeping

* Greenhouses, cloches, cold frames, and shade-houses: design and use

* Weed identification and management

* Composting & soil amending

* Dry-farming and water-wise irrigation

* Cover-cropping

* Plant communication

* Farm budgeting & financial sustainability in changing economic times

Methodology: Lecture + Lab. Instruction will feature hands-on agricultural work, lecture, and Q&A led by Farmer K and other local growers. Students will receive hand-outs each session, for assembling a 3-ring binder they can reference and share.

Students will be offered fresh produce and staple crops, seed for their own projects, and more.

Children of students are very welcome to attend.

The farm will provide all tools and implements. Students provide their own work gloves and lunch and are encouraged to eat a big breakfast.

Recommended pre-reading: "Growing Vegetables West of the Cascades" by Steve Solomon, "The Maritime Northwest Garden Guide" from Seattle Tilth, "One Straw Revolution" by Masanobu Fukuoka

PRICING: $650, with at least half paid up front; Work-trade offered for discounted tuition or full scholarship. Enrollment is limited to no more than 10 students.

Nobody ended up showing any interest in paying and only two people (a couple) inquired about work-trade. A week later, I made this follow-up comment about the idea:

Date: Thu, 20 May 2010 10:17:32 -0700
Subject: [sunrootgardens] Saturday Work Party in Hawthorne District

...By the way, any day you come out to help, you are being offered exactly what was pitched in the course that the farm is offering on alternate Saturdays this year. That is to say, you can skip paying for the course and just come to help if you'd like. Setting up a pay-course was simply done because many people around me believe that's what many other people might want. We will see!

For those playing along at home, I believe that charging-for-a-course might tip Sunroot into the "sold-out" category, which awards your team 50 pts as we move into the next-to-last round of this competition. Wagering is encouraged, but know yer bookie.

— "It's not about being pure. It's not about keeping your hands clean or avoiding guilt. Imagine birds living in a forest. Humans come and cut the forest down and build barns and plant crops. If some birds are able to live in the barns, or eat the crops, they don't say, 'I'm not going to live in the barn—that's cheating,' or 'I'm not going to eat the crops, because then I'm just part of the system.' Of all the species on Earth, only humans are that stupid."
-from "How to Drop Out" by Ran Prieur (http://ranprieur.com/)

The lack of helpers in 2010 became a common subject and then running joke amongst the farmers working together that season:

Date: Sun, 30 May 2010 21:52:37 -0700
Subject: [sunrootgardens] Monday workparty

...We farmers often find ourselves out in the field, under the great sky, with cat friends and the caws of crows, having a total blast, and chuckling over the lack of other folks—you know, the ones who say they "want to get their hands dirty" and then never show up. "Nobody knows how much fun we're having!" we have heard ourselves remarking lately, and then we shrug it off. People wake up to what they wake up to when they wake up to it, and that's about all you can say! In the meantime, we're out there, being free and having fun, figuring it all out together....

Of course even that comment was a "sell," bragging that the farmers were "having a total blast." We sincerely wanted to share our good times with other people and we were disappointed that there were so few takers. The people who did show up found a fun scene, as promised, and some even joined in.

But by August of 2010, I was done with helpers. From that point forward, Clarabelle and I would only work with close friends such as Angel. We had also been joined by a kitten at that point, a grey tom we named, "Fugz," who ended up growing into a big, burly farm-cat.

Date: Fri, 27 Aug 2010 22:42:01 -0700 (PDT)
Subject: [sunrootgardens] Helping with Winter-Crop planting

This is the time of year to be seeding many vegetables that are harvested over the winter. The amazing array of produce that can be picked fresh from Dec-March is a quality of this bioregion that makes it special not just in the U.S., but in the world. 365-days of harvests will be increasingly appreciated over time, as economic and social conditions continue to change during this empire's fall.

The knowledge about winter-harvested vegetables is not commonly held by the local population, and even most farmers in the area are ignorant. And yet, this food is essential. More so than the summer-harvested vegetables, which are everywhere. I cannot stress too much how important this point is.

The farmers at Velocifeed and Sidewalk's End will be planting up winter-harvested veggies for the next few weeks. This is an excellent opportunity to learn about these vital crops.

Please contact these farmers to find out how you can help.

For my part, I am no longer taking helpers, so there is no need to contact me about that. There have been innumerable opportunities over the last few seasons for people to take advantage of my experience and knowledge of Cascadian Agriculture, and now those days are over. I expect to continue advising the urban farmers listed above, as they all have a level of seriousness that makes them worth talking to. I no longer have patience for newbies.

I am not the first person to remark on the inability of most Americans to be good students.

The din of the culture's ego-driven engine drowns out the good senses required to learn. Pride and attachments prevent most people from being able to approach an educational opportunity with the openness required to pick anything up. Here in PC Portland, personal dogma has been the most frequent flavor. The drive to categorize every activity as good or bad—for the environment or for the government or for the ego-self, etc.—has created a confined space in which facts are ignored, freedom is squashed, and feelings are fetishized. This maze of traps in the mind is mistaken for "The Real World," although it is entirely abstract. Meanwhile, the *real* "Real World"—the world of experiencing life in real-time, through the senses—is artfully dodged, although it is entirely material.

Most people who have shown up because they claim they want to learn about farming have spent their time with me from the viewpoint of being in their labyrinth, and they have been unable to not only step out of it, but to even acknowledge that it exists. My own experience and knowledge of these things is direct and personal, and shadows still flit daily across my consciousness, which is still not free of obstructions. So, I ain't saying "I'm perfect." Although, of course, I am, as is everyone who is reading this.

"Perfect" in the sense of a "Perfect Circle": whole, complete, entire. Not "perfect" like people use the word, as in "really really good." "Perfect" actually exists outside the bounds of any good/bad dichotomy scale. As does what is actually "Love," and not what people call "love." (But that's for a future essay.) People enter the gardens but since they never leave their labyrinth, they never actually arrive.

Being that we are all whole, complete, and entire at all times, no matter what we are doing, and regardless of how any action or thought can be categorized as being "good" or "bad," there is nothing to be afraid of, including that which is called "death." It is only our ego-selves that suffer. The rest of the holistic being includes a body with reflexes and automatic processes, a brain with unparalleled cognizing processes, and a set of senses that detect the energies around us, not limited to their physical manifestations such as taste, smell, touch, etc. This holistic being is merely experiencing life, with no judgment or search for "meaning." The ego-I, however, is the place from which most people say "I," which is ignoring the whole for the sake of one part. "Not seeing the forest for the trees." To be a good student, farmer, lover, or anything else, one must step out of ego, and live from an awareness that ego is merely one small part of the self, and that identifying oneself primarily with the ego is the source of all suffering. "Pain" is different, and can be felt throughout the holistic self. "Pain" is the result of functioning senses that are detecting something harmful to the self, including but not limited to the physical body. "Pain" is essential to survival.

Suffering is not.

So please, try to approach these farmers from a place of receptivity. If you are actually there to learn, then you should be mostly listening, not talking. Mostly picking up and taking in, not building up and pushing out. Open, not closed. The process of attempting this approach will serve you and the farming efforts the most.

And Don't Worry—J.P. and the Sidewalk's End crew won't be dishing you out this kind of flak!

In four seasons, I had gone from "c'mon out!" to "I'm done with y'all." Looking back, I can perceive the larger social picture in which a door that was flung wide in 2008—by the media and the moment—had slowly been creaking shut over the years that followed. I had experienced this syndrome and its accompanying heart-ache in other places and times with other projects and relationships, so even though I was not fully conscious of its shape and stage at that time, my gut told me that I had been right to call that season as my last.

07.30.2010: "Sunroot Gardens up for grabs"

In July, 2010, I announced the pending dissolution of the Sunroot Gardens operation, and outlined everything that would be made available, from gardens to tools to contacts. This is a long email because it includes, a) the list of everything that was being given away, which is long, and b) many of my conclusions about urban farming and its prospects. I have included it here to show just how big the operation had gotten and because it speaks well about what I had learned from careful observation and thoughtful reflection.

Date: Fri, 30 Jul 2010 22:48:02 -0700
Subject: [sunrootgardens] Farmer K departing; Sunroot Gardens up for grabs

Greetings from Farmer K:

I will be leaving the city of Portland sometime this Fall/Early Winter, for parts South. So now, today, begins the breaking up and divvying up of Sunroot Gardens to anyone who is interested in pursuing urban farming in this area. Listed below are the plots and other resources being made available.

Note to Land-lenders: Cultivation and harvesting will continue on your properties, including the planting of winter-harvested vegetables, unless you would prefer otherwise. During this period, please decide whether you would like your yard to continue being used for urban agriculture; if so, there will be new people for you to meet and make arrangements with. You can also expect an increase in activity on your property, as new people show up for the take-overs. Admittedly, Sunroot has been spread a little thin this season, and by breaking up the operation into smaller pieces, more time and attention will be given to each plot. (The ongoing loss of half of Sunroot's staff when Angel injured his knee two months ago has arguably had a marked effect on plot maintenance.)

SUNROOT RESOURCES AVAILABLE TO URBAN FARMERS:

The following PLOTS, all in Southeast:

"The Firepit Clutch":

* The Firepit Garden (Double lot, 1/2 perennialized, incl. medicinal herbs, French raspberry patch, and fruit trees. Other 1/2 self-seeding annual vegetable beds. With greenhouse/tool-shed, bikes & carts, some tools, and more, see below); since July 2007

* Ninja (Vegetable beds in front, back & side, with two large cold frames; strip of medicinals); since 2008

* Tic Tac Toe (Straddles two backyards, includes giant three-binner composter); since 2008

* Logan (Small side yard filled with perennial culinary+medicinal herbs); since 2008

* Kaia (Front & Side yard of annual vegetable beds, some culinary+medicinals); since 2007

* Baby & Ollie's (Back & Side yard of annual vegetables); side since 2008, with back new in 2010

* JJC (Front yard of annual vegetables); new in 2010

Other Vegetable Gardens:

* Cleo (Front and side yard incl. right-of-way of unpaved street; annual vegetables); new in 2010

* Mall56 (Large front yard, with small back a possibility; annual vegetables, many culinary+medicinal herbs; a Primo Spot); since 2008

* Gypsy (Front and side yard, annual vegetables); since 2009

* Tolman (1/2 of an empty lot, plus unpaved right-of-way; annual vegetables); since 2008

* Roly-Poly Sneezestress (double & 1/2 lot, annual vegetables); since 2008

Medicinal Herb Gardens:

* Cora—"My Old Favorite" (unpaved right-of-way); since 2006

* Cabbit—"My New Favorite" (unpaved right-of-way); since 2008

If you are interested in any of these plots, see "HOW YOU CAN GET PLOTS," below.

Also available:

*** "SUNROOT GARDENS" BUSINESS ENTITY ***

For sale, for $1 OBO ("Or Best Offer"), bidding is open until 13 August. The enterprise includes:

* established name, a media darling

* CSA Customers: some number of current subscribers will likely be interested in produce shares post-October for winter season, and/or full 2011 shares; could be enough for a full plate with no need to recruit more for the season

* The Firepit Clutch of 7 gardens, pending continuance with land-lenders

* Introductions to local businesses where barter/exchange has historically occurred

* Introductions to Friends-Of-The-Farm (FOTFs)

* Introduction to space sponsors of Sunday Southeast Urban Farmers' Market at 43rd & Hawthorne; handover of The Key To The Gate would likely result

* Distribution point for CSA shares at Firepit Garden (pending agreement with The House)

* Some hand tools, mostly the old ones; enough for a first-year "starter set"

* Some powered tools, i.e., weed whackers or possibly a tiller, in various states of workability/fix-ability

* Bikes and bike carts, customized for urban farming

* Greenhouse space with adjoining insulated "farm office," and a tool room (aka "The Fort")

* The cold frame farm at the Firepit, for gardening in the French style, "under glass"

* At the Firepit, a collection of random raw materials useful to farming, incl. lumber, PVC, poultry fencing, cordage, bits of hardware, hoses, pots, etc., etc.

* Website with hosting (pending continuance of donator trade); includes "blog"

* Possibly "the Farm Phone," a cellie with a widely published number, through which many inquiries are made ("leads"). Continued use of the Farm Phone is pending agreement with the holders of

the hosting family plan.

Bidding is open until 13 Aug. You may text or call with offers. 503.686.5= 557

*** SEEDS ***

But please, to current/future urban farmers only, not home gardeners.

Much of the Sunroot collection is saved seed, featuring some original crosses, and geared toward production scale and scrappy survivability. The seeds will be given only on pledge to do one's best to continue growing out some for more seed, in serious amounts, for future farm use and for distribution to the local community (including home gardeners). Those interested in seed will be required to help with this year's seed-saving operations, currently under way in earnest.

*** THE MEDICINAL GARDENS ***

Besides Cabbit and Cora, which are predominantly medicinal herbs, the Firepit Garden is also a treasure trove of valuable plant allies. I have focused on herbs for damp winters, women's reproductive health, and the nerves. How To Stay Away From The Doctor and The Abortionist and The Pharmacist (including the happy-pill peddlers). The beds at these gardens are filled with more booty than I could easily list here; only a tour will do.

A cooperative effort among urban herbalists could tend to the medicinal beds. This would include weeding and cultivation, harvesting and processing, and dividing and increasing, "Mother plants" of several-seasons age are included, appropriate for propagation. This is work easily done by bicycle, occurring year round, much of it during the cooler, wetter months, when it's really nice to be outside rainbow-spotting.

For-Profit medicinal operations are not welcome to apply. If you are requiring $ from people for the medicine you are making, keep to your yourself. These plants have been tended with the knowledge that they are to be given freely and unconditionally to whomever needs them when they need them. That being said, herbalists have traditionally accepted gifts—including currency— from the grateful users of medicinal herbs, but only without expectation of such. Being a necessity, our medicinal plant allies do not belong in the world of commodities.

*** HOW YOU CAN GET PLOTS ***

By coming and working on them, ASAP. This is the time to plant winter-harvest staple vegetables such as carrots, rutabagas, kohlrabi, and cabbages, and soon peas. In a couple moons it will be time to seed turnips, spinach, radishes, corn salad, etc. These are the crops that YOU will have to harvest and distribute, starting post-October when the current CSA season ends and the next begins. We will be making agreements with you about how much we will be road-tripping back to harvest during the cold months for our own use Down South.

There will be bounty for everyone if you are willing to jump into the game now or very soon.

Once you step in on a garden or gardens (and land-lender understandings have been found), it will be "your" garden as much as "ours." We are happy to step out of the lead role on most of these plots. But this gift will happen only for those people who are present to receive it. "Wanting To Get Into Urban Farming" doesn't count for anything. "Doing Urban Farming" is what's going on here. This would be Real Life On The Ground Experience such is not available from anyone else in the area.

Garden pass-overs will feature oral histories including past plantings and results, so that reasonable choices can be made re. rotation choices, soil amending, cultivation methodology, etc. This will be part of the Field Work; there is no Lecture component to these studies.

The "Right of First Refusal" for the Firepit Clutch of seven gardens goes to the Sunroot Gardens Business Entity and The Herbalists. If it makes the most sense to those invested, individual gardens could have individual farmers working cooperatively under that aegis. The Firepit Garden itself probably requires 5 people to replace my efforts, so people should get together on this one.

*** SO SHOULD YOU DO THIS? ***

For those choosing to stay in the city for the time being, the opportunity to take over these gardens is remarkable. I mean, here I am remarking on it: You've got garden plots that will be going into their second, third, fourth, fifth (and one of them a sixth) season. All the hard work of breaking ground and fighting the lawn and blackberries has already happened. Much amending, some of it quite generous, has been applied to the soils, some of slow-release that will still be performing for you next year and beyond. There are individual beds in certain gardens that self-sow so well that they only need to be weeded and watered. The Firepit features lots of ground like this, where one could probably go a whole 'nother season without planting anything new, and just select to the best of the varied volunteers.

I have found urban farming to be an excellent way to live lighter, and with much less dependence on the machinery of the culture around us. At this time of year, I have hardly any need for $. The produce can be exchanged for many, many goods and services around town.

That being said, I must add that—in the six seasons I have been exploring urban agriculture—I am forced to conclude that there is No Market for local produce of this quality. People don't yet understand the incomparable value of produce that is so local and so fresh and grown with such intimate interplay.

This is exemplified (among many other things) by the experience of the Southeast Urban Farmers' Market that we set up last summer, and have continued—so far—into a second season. Yep, Hawthorne in the 40's, fee-free, is an excellent location. Yep, everybody says Portlanders are local-food-crazy. Yep, urban farming "is really taking off," yet there is often only one vendor. Last Sunday, we were set up for three hours and made three dollars total. The current psychology of the marketplace is that if there aren't many vendors, it must not be worth it to stop in. Farmers don't want to commit time to a market with no customers. This psychology continues to the market table, which must be "fluffed up" to look "bountiful" and have many "selections" or no one will want to look. Finally, almost no one is actually interested in What's-In-Season; they come with a shopping list, and are hesitant to try the unfamiliar. This vicious circle will not be broken until a cultural shift occurs, and people recognize the *real* real world, in which local food is understood as Essential.

In all likelihood, such a shift will not happen without some kind of "Empty Shelves" event, and then you won't be able to grow *enough* produce. I'll be sitting that one out, thank you, someplace where I might not even hear about it!

If you have any expectation of supporting a conventional, consumer-driven lifestyle that includes a mortgage, health insurance, evening entertainment out-on-the-town, or any of the other trappings of middle-class life, urban farming is not for you.

If you are serious about growing food and are willing to jump into it, seeing that you are being offered an incredible head-start, then give me a call. Introduce yourself as an Urban Farmer, because the moment that you want to be one, you will be one. You will be able to free yourself of living in the world of ideas, dreams, and abstractions about "urban farming," and actually just do it. You will not be learning a Methodology. You will be discovering how to cultivate your own senses and awareness, which can open into a whole new world of satisfaction and delight—with the best damn food around!

You can jump in immediately if you'd like. We have Harvests happening every Sunday, Tuesday, Thursday, and Friday, which entails picking and transporting. We have a market to be covered on Sundays. We have Seed Saving hitting its busy season. We have irrigation to keep up on. We have winter-harvested veggies to sow. The gardens are all over Southeast. You're sure to live within workable distance of one or more of them.

Over the next three to six moons, you will have winter cover-crops to sow, garlic to plant, cloches to prepare, perennials to harvest and divide, roof leaks to fix, compost collection and turning, and harvesting, harvesting, harvesting. Waiting for you on the Winter Solstice will be delicious, candy-like parsnips we planted for you in April. We've been consciously preparing the way for you. And that was just by doing the best farming we could do.

But we're outta here soon, so now it's your turn !!

I received a $2 bid from another urban farming operation, Sidewalk's End, for the Sunroot Gardens business entity. However, they never paid it, so the sale was never finalized. Nonetheless, they did take over a number of gardens.

08.27.2010: "Sabotage"

Date: Fri, 27 Aug 2010 22:42:01 -0700 (PDT)
Subject: [sunrootgardens] Sabotage

Sabotage @ the Firepit Garden

About a week ago, an unknown substance was applied to much of the gardening space in the beds along the sidewalk at the Firepit. Within 5 days, nearly every plant had been killed. The plants were from too broad an array of botanical families for disease or pestilence to be the culprit. Lack of watering was not the issue, as half of the beds affected were regularly irrigated. Whatever it was killed some fennel, for chrissakes. *Nothing* kills fennel in this bioregion. Look how well it grows out of pavement cracks in notoriously toxic areas, such as along railroad tracks.

The perpetrator and motive are unknown, and there are no leads. It is quite a mystery! We plan to get samples to PSU for testing.

Before you say, "I'm sorry," let me tell you not to bother. On one level, it is merely a crop failure, and those are inevitable in farming; something I've gotten used to, at this point. Too many factors in the success of a crop are out of my hands for me to take the credit or the blame, The Weather being the biggest one.

One another level, the incident is an attack. Whether personal or not, I don't know. In that way, it is merely an expression of the energy of The Times. The energy of Our Times can be characterized as entropic, fragmenting, or with increasingly chaos. This is not Woo-Woo. Simple observation of the world reveals an environment in flux, in which established patterns are coming to the place that all patterns eventually do: an end. This is not end-of-the-world-ism, this is simply the physics of the planet.

This energy is something that is sensed by all things that have senses. Like animals who know how to high-tail it before the tsunami hits, we as human animals are picking up on this changing energy, and we are all reacting in different ways. Some are opening up, like the bud before a flower. Most are closing more, like the ostrich burying its head. These are not good or bad things; they are merely Conditions of Our Times, just as each day has Conditions of Weather.

They are nothing more than a part of living with a body.

This energy manifested itself in an attack on the Firepit Garden, and that is no surprise. Every living system acts to protect itself from attacks, and the spirit of Sunroot Gardens has very much been an aberrant mutation to the system around itself.

It doesn't belong here, and the attempt to push it out is detectable in many other manifestations. Again, this ain't woo-woo, it's just the physics of living systems that are seeking to perpetuate themselves, which is what they all try to do.

If you are sorry because you are living in a time when food-growing comes under attack, then you need to look at things as clearly and closely as you can, and try to detect how you yourself perpetuate that system by letting it set up shop in your own head. This process of close observance is nothing to shy away from. It has been claimed that finding such awareness of your self is the only real reward of living at all.

The perpetrator or perpetrators of the sabotage were never positively identified. Looking back, I suspect Augusta and Lairy, or possibly their henchman, Delta. Delta was a vile manipulator and pathological liar who had attached himself to Augusta and Lairy.

Unmentioned in this email is that, when I first observed the dying plants, I knew at once that it was sabotage because I could see, with my mind's eye, the mark left by *malice*. Malice is a purely human product. The lion that takes down a gazelle is not malicious, nor the coyote that nabs a rabbit, nor the spider that consumes the juices of a still-living fly. Not even the house-cat that bats around its prey is acting from malice; we call their behavior "toying with" but that is only our own malicious nature attempting to justify itself by imprinting it on the feline.

We have, all of us, felt malice at some point in our lives, if only just for a moment. That is the inevitable result of being raised by a malicious society. But so we are able to recognize it when we see it, a perception we would retain even if we successfully rooted all malice from our own beings. In the case of the sabotage at the Firepit Garden, I knew what I saw, and I saw what I wanted to expel from myself.

Wasted Seed and Murdered Bees

"Lobelia" owned a house across the street from the Firepit. Her property was packed with an impressive variety of plants; it was a beautiful garden, well-known in the neighborhood and deservedly so, but small. When I first took over the Firepit space in 2007, she approached me about cooperating on it but a little voice inside me told me to keep her at arm's length so I declined her offer, but with tactful care. I didn't want to piss off a neighbor.

As time went on, her prickly personality revealed itself, and this led to our secret nickname for her: "Lobelia," named for Lobelia Sackville-Baggins, a character in J. R. R. Tolkien's legendarium notorious for her covetous nature and sour demeanor.

I did, however, enjoy actively friendly relations with members of Lobelia's household throughout the years I lived and worked at the Firepit. Her house-mates were a constantly revolving cast of characters culled from Portland's New Age circles and all of them were warm in their interactions. Many of them were a little (or more than a little) vacuous, but a few had a genuine interest in the importance of personal ("spiritual") work, so meaningful interactions were enjoyed from time to time. But few people stayed in that house for long. Lobelia was a queen-bee who didn't want to share her throne or even have it approached too closely.

Lobelia kept a number of bee hives. At some point, she asked if she could place one in the Firepit Garden. I enthusiastically accepted the offer. Not only would bees be useful residents in a space where so many plants needed pollinating, but I also loved honey and she promised to share some with me in return for the favor.

That hive quickly became a popular attraction in the garden, piquing the curiosity of most visitors. Children especially were fascinated by the constant stream of insects flying in and out of the tiny entrance. Unlike

Lobelia, the bees were gentle, never looked for a fight, and only stung someone once. (In that single instance, it was a sketchy individual who, on some level, was asking for it, and I didn't feel bad for him.)

In the Spring of 2010, a swarm broke off from the hive and gathered in a swarm on a small branch on one of the apple trees. Lobelia had an extra empty hive so we quickly brought it over, found a good location for it, and set a shallow dish of water with crushed Lemon Balm foliage inside it. Then we carefully cut the branch from the tree, placed it in the box, and put the cover on. Standing back, we crossed our fingers; bee-keepers have lots of stories about captured hives that pick up and leave, so success was not assured. As it turned out, the bees decided to stay, though how that process works, and if it can be called a "decision," I don't know. Bee consciousness is quite foreign to a human conscious-ness, at least to mine.

During the winter of 2009-2010, I noticed that a patch of mixed let-tuce in a raised bed in Lobelia's side-yard was surviving the cold quite handily. Lettuce is touch-and-go over-wintering in Cascadia, so I asked her if it could be left for a seed crop that I could harvest. In return, I of-fered her household what any other landlending household got, which was a CSA share. This was quite a generous offer since all of the other landlenders were giving us a lot more space than just a single bed. In fact, in the past we had turned down offers of single beds as being too lit-tle to be worth it. But in this case, I felt that the value of that lettuce's cold-hardy genetics was great enough to warrant the exchange. I was aware that Lobelia's sense of self-entitlement prevented her from realiz-ing what a good deal I was offering her (especially since her household of 4-5 adults would take as much produce as two smaller households) so I kept my mouth shut.

Now, it must be said that saving seed from lettuce requires more pa-tience and attention than many other vegetables. For one thing, *Lactuca sativa* takes its time to flower, not bolting and blooming until mid to late Summer, depending on the variety and when it was started. This lengthy maturation period also means that the seed crop is in danger if the fall rains return early that year, which can prevent the seeds from drying properly. So if you're growing lettuce for seed, you must sow it as early in the Spring as possible. Over-wintered lettuce, like in Lobelia's bed, is the best; it will go to seed sooner and give you a better chance of success.

Additionally, lettuce seed is not ready all at once. Each lettuce plant produces a few hundred flowers, but only a few open each day over the course of a several-to-many weeks. Each individual flower closes—never to open again—after a just a few hours. As each seed finishes, it is ready to be caught on the wind and blown away, carried by its Dandelion-like tuft. Once the first wave of seeds ripens and dries, you need to be out there every few days harvesting it as it finishes, for several weeks.

Having lost so many seed crops in the past, we knew we needed to do whatever we could to lower our risk of losing this one, so we posted a clearly printed and worded sign in the bed, right in there with the plants, but still plainly legible. We knew we couldn't count on having a clear arrangement with the landlender; someone else could take out the patch once it started to look "weedy." The sign instructed anyone who read it to *not* harm or remove the plants and explained that we were seed-saving. Short of building an electrified cage around the bed, we didn't know what else we could do.

With the summer's heat, the patch of lettuce started bolting. To me, bolting lettuce plants resemble far-out Far Eastern architecture: spiraling tapered towers of layered wavy-margined cornices crowned with a halo of tear-drop buds. Sitting and staring at them, I had visions where the plants were many, many stories high and I was walking among them, marveling at such an exotic, psychedelic, city.

But the plants weren't giant, and it would only take a minute for someone to "clean up" the bed and destroy them, after so many months of waiting and so much produce given to the household. At the time of year in Cascadia when cultivated Lettuce buds out, Wild Lettuce (*Lactuca serriola*) is also in the flowering stage and is nearly identical in appearance. It is common—and commonly yanked. As our precious *L. sativa* seed crop started flowering, *L. serriola*s were being grabbed by gardening gloves and hitting compost bins all over town. I made sure that the "Please Don't Pull!" sign wasn't getting lost in the patch and kept a close eye.

Rain came! and went away without doing harm. Petals unfurled. Bees, wasps and butterflies visited. Blossoms closed and—gradually—Dandelion-like tufts emerged from each bud as the green color drained out of the spent buds. One day, finally, the first wave of seeds was ready. Clarabelle joined me on the project and we gently bent the heads inside paper grocery bags, taking great care not to break or crease the stems since each one still had many buds yet to open. Finished seeds fell right out, making a satisfying rattle as they hit the sides and bottom of the bag. Painstakingly, trying not to knock seeds off just by touching the plants, we worked our way around the patch, letting all the varieties mix. After all, it was a nice blend: Romaines, Butter-heads and Oaks in various Greens with a hint of Ruby.

Over the course of the next ten days we returned to the patch twice more and collected about a third of the potential harvest. No more precipitation threatened—it would have been uncharacteristically early if it had—and we seemed right on track. Someone from Ruby's household came to the CSA pick-up that night, as they had been doing weekly all summer.

Then, on the fourth visit, the plants were gone. All that remained in the bed was the "Please Don't Pull" sign. We were shocked. And more than a little disappointed. And just plain angry. We took a look around and found the mangled plants in a nearby compost pile. Digging through other detritus, we removed as much of their remains as we could find and brought them back to the Firepit where we hung them from the rafters in the shaded part of the greenhouse over a tarp.

The loss was significant: The rough treatment of the plants had shaken off lots of seed that was gone forever. Many seeds were still too immature to ever ripen. Some unopened flowers remained. A few seeds were far enough along that they finished successfully on the plants. In the end, the total take was well less than half of what we had expected. I didn't say anything to Lobelia. I didn't want to talk to her when I was angry about it, and I knew her well enough to be assured that I would only get excuses and no apology.

A few days later, I was working out front in the driveway with J.P.. We had a tarp spread out, some buckets, and a very large pile of seeded radish plants that we were processing. That evening's weather forecast was calling for a high chance of rain, so we needed to break the plants down enough to get them undercover in the greenhouse. We were racing to beat the light; the sun had just touched the horizon and soon it would be too dark to see. One of the members of Lobelia's household biked by on her way home and stopped to say hello. I can't remember exactly how it came up, but it was well known at the time that I was looking for new stewards for all my gardens, including the Firepit, and it was also well known that Lobelia wanted it. I said to Lobelia's house-mate that, because of recent events, I had advised Lady Quince not to allow Lobelia to have a part in the place when I moved on. The house-mate's eyes got wide and she mumbled a quick goodbye and took off. I soon regretted that I had said anything.

Not two minutes later, Lobelia came stomping over, steaming mad, with a face as sour as her namesake's. She demanded to talk *now*. I told her I was too busy to stop what I was doing, and kept processing. The shadows were lengthening, and the pile of bolted radishes was still high. She wanted to know why she was being cut out of the Firepit. I told her it was because she had yanked the lettuce seed crop before it was finished. She insisted it had been done and that there hadn't been any more seeds left. I said that, no, it had not been done. She said that, yes, it was—she had checked first. I sighed, dropped my task, went back to the green-house and brought out the lettuce branches. I held them up in front of her and pointed out the seeds. "See. There *were* seeds left. Here they are. And here's flowers that never even made seeds. You pulled them too early."

Without skipping a beat or even blinking, she changed the subject, but not her tone, and started in on something new, but as if it had been the topic the whole time. I stared for a second—I was quite taken aback by her effortless switch in gears with no acknowledgment whatsoever that I had just proven her wrong.

The sun was well below the horizon. I had very little light left. I told her I was busy and that she needed to go away. She refused. I said it again, more firmly. She refused again. I couldn't believe it. Wasn't the re-quest to be left alone sacrosanct? Weren't you playing with fire to ignore it? I raised my voice but not to the level of shouting, "Get out of here! Leave me alone!" "No!" she said, and would have literally dug in her heels if she hadn't been standing on a concrete sidewalk. Then I did shout. "Leave! Me! Alone!" and spat on the ground between us.

Now it was her turn to be taken aback. Her eyes got huge. Her mouth fell agape. And then, finally, she turned around and left.

J. P. had witnessed the entire exchange an seemed a little taken aback too, though he usually presented himself as pretty unflappable. "People just don't know how to have enemies anymore," I growled. There was a pause. Then he smiled. We both chuckled, and returned to the radish seeds. We ended up getting the task done, but just barely.

The next day I was away from the Firepit, out at some other gardens, and when I returned both the bee-hives were gone. I knew Lobelia had taken them and I was pissed off. The bees didn't have anything to do with our conflict and you weren't supposed to move hives that far all at once, and certainly not when it was light. But she had done it anyway. In the days that followed, dozens and dozens of bees showed up at the site of their former home and buzzed around with obvious confusion. A plate happened to be sitting on the ground below the spot where their entrance had been, and soon it had filled up with dead bees.

Angel visited the garden, saw the dead bees, and was very angry. He was especially upset about the second swarm, which—he correctly pointed out—had *chosen* to be there, in the Firepit garden. What right did she have to move them? She had nothing to do with that hive except that she happened to own the box. Clearly, care or concern for the bees was not foremost in her mind when she moved them. Revenge was her motive. Angel knew all of this and was disgusted. The bees had been the only casualties of her action. I, personally, had taken no hit at all from the event, other than missing them. Tolkien's Lobelia had never been that destructive.

I heard through the grapevine that Lobelia had characterized my behavior during our exchange as "violent." That really peeved me. I hated how the liberals in Portland threw that word around in a way that only people who have never experienced real violence can. I had often wished that these people could be sent to Afghanistan or Iraq or Gaza for a month and see what violence actually is.

So, I took the plate of dead bees to her house and left it on her porch in front of her Buddha statue with a note saying, "*This* is violence. Killing other living creatures for the sake of your own ego."

After that, she left me alone. But Angel still gets angry if you bring up those bees. As well as he should.

09.18.2010: "2010 Farm Year Notes"

Date: Sat, 18 Sep 2010 14:26:53 -0700
Subject: [sunrootgardens] 2010 Farm Year notes

...What a fucking year this has been! In my time farming in the City of Roses, since 2005—for seven seasons now—I have never seen a worse farming year. The themes so far have been Pestilence, Drought, and now—Monsoon.

Pestilence: it was the worst year for slugs I have ever experienced. We have never had to reseed or re-transplant so many things. The tobacco crop suffered a 90% failure rate to slugs. Beets, carrots, and lettuce would come up and then disappear.

The Rains didn't let up until the end of June. The cool wet weather got the beans, tomatoes, peppers, eggplants, squash, and corn off to late starts. Though a few patches here and there flourished for a time, many never got going. At the Firepit Garden, fifteen (15) different bean plantings, ranging from very early to a little late, all failed. It's true that five of those were in the patch that was poisoned by a saboteur, but this too is a crop failure. At least one pole bean planting has *still* not started to bear.

The Rains were followed by about 9 weeks of Drought, with barely a drop falling from the sky. This condition stressed or killed many more crops. In the "distributed network" that we urban farmers are saddled with, keeping up on irrigation everywhere proved impossible. Some plots were dry-farmed by default. Some crops did well during these circumstances, but many more stopped or never took off at all.

And now, as if all of this hasn't been enough, we had October Rains arriving in early September. This is the death knell of the summer squash, the stall-out of the beans, and the dashing of any hopes of big juicy ripe tomatoes.

The cabbages that survived the slugs have been doing fabulously. This has indeed been "A Cabbage Year" [as predicted by Clarabelle], and some of those are on the way soon.

These early rains on still-warm soil are also providing the perfect conditions for late summer / early fall greens and root crops, and for bringing up and growing out winter carrots. Those seedlings that don't get munched by the happily returning Slugs will give us some delicious produce in October. We have tons of spinach and mustard greens coming up everywhere, as well as turnips for the cold months.

This bioregion remains one of the best food-growing places in the world. The only challenge is for consumers to adjust their expectations to match the supply. Most farmers' market shoppers, local restaurants, and CSA subscribers still have "Shopping Lists" when they go seeking local produce. Most local farmers are still trying to cater to these expectations, rather than educating the consumers of the benefits and limits of this area.

I remain appreciative of the Sunroot CSA subscribers, who are exceptions to this rule, and who it has been my great pleasure to offer produce to. My appreciation of my Sunroot customers grew this year, as I often sat in on the Riverhouse CSA produce distribution, and saw what kind of customers most other farmers have. The privileged moneyed liberals of Sellwood are generally a pain in the ass, with little-to-no appreciation of the labor of farming, and the limitations of the area. Some of the Riverhouse customers are not like this, it's true, but Oh Heavens, the ones who are! Yes, I am indeed very grateful that I kept Sunroot exclusive for the last two years, only letting in a few members of the general public who did not have some connection already.

The market for local urban produce remains tiny. In the seven years of farming, I have seen more empty lots go to condos and narrow-houses than to gardens. I have seen more lawns re-sodded than taken out for food. I have seen the self-congratulatory "We're a Foodie Town" talk of the local press become more and more of a pretense. Despite the increasing challenges of These Times, and the growing need for local agriculture, I am not seeing any real progress. The only reason that I myself have been able to support myself in these endeavors over this time is by not having to pay rent for over two years (and by having below-market rent offered by landlords before that), and by drawing on the resources of communities that I had connected with previously, such as the political radicals and the Co-op crowd. Also, we cannot leave out the undeniable effects of receiving so much local press—something like a dozen media hits since 2008. This surreal notoriety has proven invaluable for the farming efforts, especially since I never took it personally, and merely used it to provide for the project.

So as this season comes to a close, I am filled with a sense of relief. It has been an enjoyable ride for sure, and my own holistic growth has been the best part for me. But the number of challenges working against agriculture in the city have finally piled up high enough to overtake the rewards, and it is time for me to stop. Like a plant that has outgrown its pot and needs to be put in the ground, it is time for me to say "Thank You" & "See You" and move along.

10.10.2010: "Staple Crops Report 2010"

Date: Sun, 10 Oct 2010 16:58:40 -0700
Subject: [sunrootgardens] Staple Crops Report 2010

Here is the long and the short of this year's Staple Crops Project.

The short of it: There ain't nuthin' to distribute. Nuthin'.

The long of it:

1) MISSED SPRING PLANTING WINDOW:

The planting year began with an atypically dry spring, starting in late winter. From January through the end of March, a beautiful tilling window existed, during which conditions were flawless for working the soil and getting in crops. However, we experienced complete equipment failure during this time, at one point having not one but four machines for tilling that were broken. These were:

* the Massey Ferguson tractor (busted hydraulic system, the repair of which would have required several hundred dollars, plus the ability to transport it back and forth to the shop),

* a mini-tractor (acquired at a cost of $700 and which tilled exactly 10 feet of soil before breaking; two attempted repairs by skilled persons did not fix it), and

* two Troybilt tillers (one leaking fluids at a flow that would have polluted the soil, the other simply not working; one was repaired for a cost of $700, but ceased effective operation after one job).

Hence the optimum sowing window was missed for most grass-grain crops, including wheat, oats, and barley, for which we had purchased seed. Torrential rains followed, during April and May and

June, causing us to miss any more opportunities for land-prep.

2) FUTZED-UP MILLET PLANTING:

We met a sister with 10 acres that she owns in Monmouth, and we laid plans to sow millet and buckwheat. Having no equipment of our own to till out the seven-year old perennial rye grass there (which had been grown for grass seed by previous farmers), we appealed to her to help us find someone in the area we could hire to till it for us. Her uncle, who lived nearby, came out and worked four acres of it. He refused to do all ten because—even though it is her land, and she wanted all ten tilled—he believed that was all she could handle. Here we see the patriarchal bullshit that happens in biological families. Additionally, he did not till it. He disced it, and only did so once over, leaving enormous chunks of sod, barely turned over. The result was a piss-poor tilth for seeding anything, or for taking out the rye grass.

The fact of the bad job was not known until a Sunroot-organized team arrived of several cars and a few hundred dollars worth of seeds and soil amendments. Trying to make the best of this situation, the amendments and seeds were spread and raked in by hand by a dedicated crew.

The weather at this time turned wetter and cooler. The germination of the crops was spotty, and the rye grew back up almost like nothing had happened. By mid-summer it was obvious that there would be nothing to harvest, so Sunroot gave the go-ahead to sister to re-till it for another attempt at her own crops.

3) THE TOO-LATE MCNUTT GRAIN PLANTING:

Another opportunity for land offered itself in Gresham. A farmer we were then working with [Augusta] reported that there were four tilled and plantable acres out there. Though the optimum window had closed, we decided to try a last minute planting of oats and field peas, as well as a well-timed sowing of buckwheat. Upon arrival with a Sunroot-organized team of several cars and hundreds of dollars of seeds and amendments, we found not four acres available, but half an acre, and un-tilled to boot. [This, and other overestimations of land size over the season led to the catchphrase "Augusta-acres."] $100 was spent paying a neighbor farmer to work the space for us on the spot that day, since we had people and seed ready. However, he too only disced the plot, and the result was poor germination and weediness.

A farmer friend [Delta] working the spot gave us an excellent report during midsummer, delivering us harvested and cleaned samples of each of the crops, with the amount of time it had taken to process each. With this information, he asked us which would be worth it to do. The answer was, "None, really." Additionally, the bag labeled "oat" seed actually contained wheat seed, so we had no oats at all!!

More crops were harvested. They were brought to the HQ of the Westmoreland Garden Club for processing, and as far as we know that's where still are. The Westmoreland folks have since acrimoniously ceased working with the Sunroot and Riverhouse CSAs, and so these harvests have been written off as stolen. Yep, stolen crops AGAIN. (See the 2008-2009 Staple Crops report for detail on that year's theft.) What was "acrimonious" about it? Phrases such as "Fuck off and die you money grubbing whore." I got into farming in part to get away from the BS personality drama that so plagues every aspect of the conventional culture, but found myself dealing with it again this season. Lessons have been learned, and the wagons circled a little tighter.

4) THE GOATRIDGE FAMILY CLUSTERFUCK:

The Riverhouse CSA farmer we worked with this year had begun the season trying to make a relationship with Goatridge Farm in Silverton. Goatridge is a goat dairy and is located on the second-oldest land claim in Oregon. What was once a 640 acre farm had been reduced to 20+ by the family over the years in order to hold on to some part of it at all. This had been the history of most large family farms in the U.S. over the 20th Century.

At this point, the farm is owned by a non-profit land-trust with a board-of-directors, etc. The farm and goat dairy there is run by a lady we will call "R." R.'s relationship with the goats was one of the most thoughtful and attentive people-animal relationships I have ever witnessed. To call it her "treatment" of the animals does not give enough credit, as that word "treatment" suggests a master-owner set-up, which was not the vibe. Just wanting to give R. props for that.

R. also wanted to find some younger farmers (she is 60ish) to develop the place into an actual working farm that would be raising vegetables and grains for the people who live and work there, and grains for the animals. She recognized that, though she was living in the country on owned-land, that she and the farm were still entirely dependent on the city and industrialized culture, and were not truly taking care of themselves.

Enter sister farmer from Riverhouse [Clarabelle], who offered to take these pieces on, with help from her friends. Sister planted some beautiful bountiful vegetable gardens, and Sunroot became involved as this was a Staple Crops opportunity. As the season progressed, it became apparent that R. was not willing to let go of these projects (which she herself had no time and few resources for) and in her insistence to square things away in a particular personal way, planting windows started getting missed. A half-tilled field never became a fully tilled field. The area offered for crops was reduced. Tellingly, the farmhouse was not seriously picking from the vegetable gardens that had been planted, and did not deliver on promised help with cultivating.

At last, in July, with most staple crop planting windows being closed, Sunroot ceased its relationship with Goatridge. The final straw was R.'s demands that the people working on the farm live by a set of family-values without which she believed no working farm could happen. I had helped Riverhouse sister with the vegetable plots in order to make relationship with Goatridge and R., and had done my very best farming for them while I was out there. So—in a lesson I had already learned many times in the city doing urban farming—I found out once again that simply offering my best farming practices was not enough; no, I also needed to subscribe to a certain set of intellectual and/or religious beliefs (what's the difference?) in order to participate. This I refused to do. Here we see the patriarchal bullshit that happens in intentional communities.

5) STOLEN CROPS:

We planted a small amount of triticale (a wheat/rye cross) and barley at J. P.'s Alleisdair plot in outer-SE. These crops did fairly well for their late planting. However, the bigger of the two—the barley crop—was harvested and processed by the Westmoreland folks, and is considered stolen as well. The triticale patch was merely a test plot and yielded less than five pounds.

CONCLUSIONS:

a) Shit Happens—the broken equipment was a bummer, and the weather uncooperative

b) People Aren't Ready—I have discovered a fundamental truth about the American society during the last six farming seasons:

*** Almost Nobody Understands the Importance of Providing Food for Ourselves ***

This vital importance—that we eat—was well understood by just about every single generation of humans in the history of the species, up until the last few decades in the industrialized west. The resulting roadblocks to attempting to farm in a serious way have hobbled my own efforts, enough that I am giving up the idea of growing produce or staple crops at a scale larger than needed by my own immediate circle (which can be counted as a single digit number, including those people in that number who are cats).

This social problem—the disconnect about the importance of food —will not be solved by techno-fixes, permaculture, or any other great sounding idea. What is needed is a shifting of gears from the luxuriously ignorant to the brutally practical. The overarching sense of Entitlement held by most Americans must be dropped. The zombie-like adherence to The System must be dissolved. Nothing less will do than a complete turning upside down and inside out of all current cultural practices.

These are not issues I can address, except as they relate to myself and to my immediate circle. I have come to see that it is virtually fruitless to try to reach people in general about these issues with my words, actions, and example. Thus, the end of this agricultural season also draws the curtain on my years-long efforts to make "positive change" in the social/civil world. The idea that "one person can make a difference" is a falsehood perpetrated by an ego-driven human mindset that is invested in avoiding any real change in any way at all.

There is no hope. My next step is to embrace this truth, and experience the liberation that this acceptance will give me: to become Hopeless Man. This work will be taking me away from the City of Roses in a few weeks. I wish good luck to everyone, including myself!

RIP Sunroot Gardens

Over the Winter of 2010-2011, the following obituary for Sunroot Gardens appeared on the front page of the farm's website:

RIP: Sunroot Gardens CSA, 2007-2010

After four years at the forefront of the urban farming movement, Sunroot Gardens has gone out of business. During that time, Sunroot farmed on approximately 42 different plots in-and-around the City of Roses, providing fresh produce and staple crops to dozens of households. Sunroot is hugely appreciative of the many people who helped make the enterprise so joyful and fruitful. It was an honor and pleasure to grow food for such delightful folks.

Sunroot Gardens was uniquely well-placed to embark on an experiment in hyper-local foodshed-building. Benefited by copious local media attention, Sunroot was offered many resources, including more garden-space than could be planted and more customer interest than could be served. Additionally, its founder was already well-known for hard work and generosity in a number of different circles, and so enjoyed a network of supporters to draw from already. As a result, Sunroot did not have to scramble for resources like many start-up operations.

Given these advantages, Sunroot Gardens was able to focus on the logistical and the practical, and to seriously explore the viability of farming in the city at this point in time and history. Questions were asked: Can urban farming provide a livelihood for farmers in the expensive city? Are people ready for their lawns to be turned under for food production? Could Portland become like Havana, growing over 50% of its produce within its own boundaries?

"Not likely" is the answer that was found.

Despite the "foodie" reputation that Portland awards itself, interest in local produce is still largely a feel-good choice of the bourgeois liberals (who are rapidly going broke) and does not represent any significant shift. Land-owners for the most part remain obsessed with the aesthetics of mown-and-edged privilege and are unwilling to accept the natural appearance of farming, which is unconcerned with tidiness. Consumers of produce—even those saying they seek the local and seasonal—still carry shopping lists, unaware that their appetites are unrealistic and born of unexamined entitlement. Would-be helpers were rarely helpful, and lacked the focus required to work or to learn. Pretensions of Portland becoming the next Havana are an empty conceit.

In short, Portland is all talk, a city full of "good ideas" but bereft of serious practices.

The barriers to vibrant urban farming are social, not logistical. There is the space, the tools, and the people to make Portland a city of gardens, feeding much of the population. But the people are not ready for that yet, not enough of them. And when the need for urban farming becomes apparent—when rising gas prices and continuing climatic change bring the global food crisis to this region—it will be too late.

Hence the shutting down of Sunroot Gardens and a move to the country.

Announcing Walking Roots Farm

The founder of Sunroot Gardens has teamed up with the founder of Sellwood's Riverhouse CSA—another urban farmer calling it quits—to form Walking Roots Farm. The three-fold focus of this new venture will be on Staple Crops, which include the grains, legumes, and oilseeds needed to sustain ourselves; Medicinals, which are essential to holistic health in any age; and Seed-Saving, which is about maintaining viable, regionally-adapted, GMO-free seed for future generations.

Walking Roots Farm is leasing multiple acres of former grass seed farm near Corvallis. This is grass-seed country, and looking out over the land is like surveying a huge lawn that goes on for hundreds and hundreds of acres. Who better to tackle such a project than former urban farmers?

2013 Season: Slight Return

The 2011 season in the country was disastrous. The Willamette Valley is a toxic place: a century of violent resource extraction has left behind clear-cut forests, dead soils, drained wetlands, poisoned rivers and ecosystems entirely out of balance. Agriculturally, the rate at which our crops suffered total failure was in excess of 60%. Socially, it was worse. The land-owner turned out to be what Holden Caulfield called a "phony." She hid her petty, spoiled nature behind a practiced smile and couldn't open her mouth without telling a lie. At some point early on, she decided we were enemies and after that she tried to make everyone who came to the farm hate me. It was an unforgivably miserable experience. That such a toxic relationship happened in such a toxic place was not surprising.

Though our contract was for two seasons, we departed in the spring of the second year. It was so bad at "Death Central" (our name for the place) that we chose to not farm at all that year rather than stick around. We brought hundreds of perennial medicinal plants back to Portland and put them in the care of Melanie and other farmers who expressed interest. We offered to share the plants and their harvests 50/50 in exchange. Then we hit the road, camping out of a Toyota truck, in search of other farmland.

For the 2013 season, we set up shop on a friend-of-a-friend's land in Sandy, Oregon, southeast of Portland. The landowner was a genuinely kind and generous woman who had compassion for our difficult experiences. She let us expand her garden from 60x60 feet to about a quarter acre, build a hoop-house and park a Streamline trailer to live in. The soil was decent and the site was clean; we had escaped the toxic valley. But there wasn't enough space for everything we wanted to do so we looked to the city again.

Our friend Comfrey had been working several plots around town, including an especially nice one that Riverhouse CSA had started in 2010. Comfrey agreed to let us farm the space for the season. It happened to be the location of many of our transplanted medicinals, so we got to take over their care again and it was just like being reunited old friends. Another previous acquaintance was still there, too: "Prosperous Calico," a superb specimen of cat who was an absolute fiend for Catnip, and after whom the garden was named. Since our last season there, another cat had moved in, "Calvin," whose lanky frame and jaunty gait presented a cartoon-like foil to Prosperous Calico's broad-shouldered build and imperious prowl. So "PC" became "PC & C" and we adored them both.

We planted beets (for Angel), cabbages, winter squash and corn. The landlender's teen-age son caused some trouble, running down cabbages with a lawnmower and knocking over a patch the corn by spitefully dragging a hose through it, but otherwise the season was uneventful. Except for the beets, none of the annual vegetables performed well but the perennial medicinals thrived, particularly the Echinacea patch, which was aglow in purple and white flowers with a halo of bees by the middle of the summer.

Back at the Firepit Garden, Lady Quince was dissatisfied with how people had treated the yard (and her) in 2011 and 2012 and knew we were looking for more space in 2013. I can't remember who made an offer first but everyone accepted the proposal that we garden the Firepit again that season. One condition I wanted (and which was immediately granted) was for total control of the space, and for the property to be marked with signs saying "Keep Out" and "No Trespassing." As before, everything was agreeable with Lady Quince. This season, though, other people were living in the greenhouse/garage, so when we were in town we slept in a room in the basement of the house, which, though dark, had a blazing fast internet connection.

Also new that year was a change in the Lady's policy regarding her house cats, whom she had always kept inside, but who were now allowed to go out. So the Firepit Garden was renamed, "Zeus & Cricket." We only saw "Firepit Cat" once. Apparently s/he ceded the garden to the newly liberated cats.

Catnip was growing everywhere. And Motherwort, Mugwort, Marshmallow, Feverfew, Fennel, Greek Mullein and Lemon Balm. There were thick patches of Parsnips, Kale, Chard, Arugula and Sunroots. A big daddy Horseradish that predated my original tenure was thoroughly entrenched. All the fruit trees were noticeably larger and needed a keen pruning.

But the property also bore damage from the 2011 and 2012 stewards, who had—apparently—followed a simple policy: to tread lightly when it came to controlling lawn-grass and Bindweed, but to abuse the medicinal plants. The culprits were people who claimed to be "doing Permaculture." They buried established perennials under sheet-mulching, crowded in new ones helter-skelter without regard to proper spacing, and smothered most of the raspberry patch under a *Hügelkultur* constructed out of plum and holly branches with fill dirt from under a walnut tree. We found valuable plants dug up and thrown into the compost pile, but some were still alive and we were able to replant them successfully.

It had taken me three seasons to extricate the lawn from that property. In the two years I was gone, it had run rampant over a third of it again. The Bindweed was back with a vengeance, climbing over everything in its path like Kudzu. When it came to plain old garden maintenance, the Permacultists had been of no use. My skepticism for that social fad rose by several more notches.

The return of the grass was so bad in a some sections that we decided to clear them completely. We hoped to plant Quinoa, Tobacco, Corn and Klip Dagga. We borrowed Comfrey's rototiller and busted out the sod.

A couple weeks and a few rains later, we returned to find the areas we had tilled covered with a living carpet of volunteer seedlings at the rate of hundreds per square foot. But my efforts at "building the seed bank" showed: 9 in every 10 plants was edible, medicinal, or both. Only 10% of what came up was a "weed": that is, had no direct human use. So the challenge was not in growing things there but in thinning them out. We had Chard and Kale, Lambsquarters and Pigweed, Tomatoes and Tomatillos, and dozens of species of medicinal herbs. In among all of this, we carved out openings for the Quinoa, Tobacco, Corn and Klip Dagga.

Our harvests were bounteous. We dried bumper crops of herbs in the greenhouse and collected pounds of seed. Two years of half-abuse/half-abandonment had harmed but not ruined the Firepit. It's fate, however, turned out to be total destruction.

In December of 2013, the landlord gave Lady Quince a three-month eviction notice. He had sold the property and when I found out the new owner was an architect, I cringed. I knew he would a.) prefer grass or "nice landscaping" to the garden, and b.) would run-rough shod over it while renovating the house, which was definitely a "fixer-upper."

Over the next month, we dug up as many sleeping perennials as we could. We had found a place to farm in 2014 on the Olympic Peninsula, near Port Townsend, and we brought them up there in a friend's 24-foot school bus. Carefully packed into boxes, they filled the entire vehicle from floor to ceiling. back to front. I made up a spreadsheet to keep track of them, and found that we were caretakers to over 3000 individual plants!

In the Spring of 2014, after Lady Quince had moved out, Angel and Clarabelle stopped by to see what had happened. The architect had gotten right to work. The house was lifted into the air and turned so the foundation could be rebuilt. The greenhouse/garage was demolished. Most of the trees were chopped down, including a 75-foot tall Doug-fir and a plum tree that produced exceedingly juicy and delicious fruit. Worst of all, the yard had been "scraped clean" with heavy machinery: there went the seed bank.

Meanwhile, on the Olympic Peninsula, we ourselves were evicted, so we never planted all those treasures from the Firepit. Instead, we gifted and sold them to gardeners and farmers in that area. In this way, the Firepit became the transformer I had envisioned it to be, as a giant "seed-bomb." The seeds just fell further away than I expected.

Afterword: Where did I go wrong?

As of this writing, October 2015, I am an ex-farmer with no plans to return to agriculture. But I feel no sense of personal failure. In general, I never did (and still don't).

Certainly, I made many farming blunders—missed planting dates, under-tended plots, badly-timed harvests—however, these are the mistakes of someone who was not only learning everything from scratch but was also taking on more than could be physically accomplished. In time, with attention and humility, the frequency and severity of such errors can be decreased, and indeed that was the trend for me, right up to my farming's conclusion on the Olympic Peninsula. The constant, active learning that I experienced, though marked by logistical mis-steps, was no failure.

Socially, conflict was a recurring theme of my farming years, as seen in some of the stories told here. There's an old saw that "it takes two to tangle" and another that "the truth is always somewhere between the two sides" but these are cliches and they are most often trotted out by people who aren't interested in examining life with honesty and intensity. Personally, I have never felt as though I had any choice except to examine life in those ways. As a result, I have long known that it only takes one person to start to fight and that though the truth *might* lie somewhere between two sides, it is often located much closer to one side than the other, and that furthermore, the perceptive eye can see that "two sides" cannot be said to exist in opposition when they are operating on totally different wavelengths. In short, in some conflicts, one party is just full of shit. Unaware of their myopia, unconscious of their ego, ignorant of their ignorance, they flail and froth purposelessly and painfully, and try to drag you into their hell with them.

At an early age, I found myself on the margins of Society looking in. I was pushed out there by those who were defining its outlines and I had no choice in the matter. At first, this was deeply unpleasant, but in time I realized I had been given a gift: never having been bestowed with the so-called "benefits" of conforming to society, I was never hypnotized by its spell. I was never taken in. Therefore, I was (and am) able to see the shape of the lies being told by most people, most of the time. Believe it or not, I have usually chosen tact and kept my observations to myself. No doubt this helps avoid interpersonal discomfort, but I am not convinced that it is ultimately the best course of action.

Fortunately for me, my alienation (and I use that word in its denotative sense—"the state or experience of being isolated from a group or an activity"—bereft of negative connotation) had the effect of awakening in me a compassion for other people and an attendant desire to devote myself to the cause of universal freedom.

"Culture is not your friend," said Terrance McKenna, and he spoke the truth. Culture, as it takes form in the USA is an oppressive force that defends itself from any threat, no matter how insignificant, by attacking ruthlessly. I am not talking about the offensives carried out on the citizenry by the corporate state (which are undeniably oppressive) but about the daily assaults of individuals on each other to "fit in," often under the banners of "family" and "friendship." Like a body's immune system fighting off an infection, Society seeks to destroy anything foreign to it. Unlike a healthy body resisting disease, however, Society strives to eliminate that which would cure it of its own fundamental illness. Seeking freedom from this cage for myself and others was (and is) my primary objective. We must find another path together; this one is leading us and the planet to ruination.

Although this quotation from Jiddu Krishnamurti already appears in this book twice, I must present it again: "It is no measure of health to be well adjusted to a profoundly sick society." The urban farming experiment, like my work with Indymedia before it, was an attempt to address this profound sickness and to offer a healing alternative. In both cases, though my efforts were small, they were met with resistance on many fronts, but primarily in the interpersonal arena. Detecting in my work and my words a foreign spirit, individuals—acting on behalf of Society as cells do for a body—sought to reject me and my message and to undermine my labors. That they usually did so unconsciously was part and parcel of Society's profound sickness: its lack of awareness.

So, yes, conflict arose when my self-aware disobedience met the culture's somnambulant obstinacy. No other outcome was possible. My motivations—concern for our collective survival, care for living things, love —were irrelevant. The drive towards assimilation brooks no denial. Knowing this, I can neither take it personally nor mete out blame individually.

None of this has made me cynical. If anything, I have become less so. As my perception of Society and its discontentment has become keener, I have experienced the fact that clear seeing is its own reward. Comprehension has been leading to acceptance (which is not to say approval).

How, then, can I view my farming career—though it was not ultimately "successful" in a conventional sense—as a failure? I cannot. The personal ("spiritual") growth I experienced as a result of challenging myself was of tremendous value. It was the most fruitful harvest of all.

Technical Details

This entire project was created with open-source software on a laptop running Linux Mint. I used LibreOffice for word processing and layout and GIMP to design the cover. Viva free software! The fonts used are: Liberation Serif, Sans and Mono for the text; Roboto Condensed for the chapter headings, table of contents items and back cover blurbs; Strenuous for the main title on the front cover; and Kenyan Coffee for the author's name on the front cover.

25554224R00130

Made in the USA
San Bernardino, CA
04 November 2015